About
Cats
and
Spooky Action

Kurt Martin

1. Introduction

The great dream of physics is a "theory of everything". Although the name might suggest this, it does not mean a theory which really explains everything. Physicists would already be happy if it were possible to unite the two great theories of the 20th century, the quantum theory and the theory of relativity, into one single theory; so far this seems to be completely impossible, as the physical conditions, under which they are valid, are much too different.

Albert Einstein first presented the theory of relativity at the beginning of the 20th century, in 1905 the special version that considered uniform movements, then in 1915 the general version that also included accelerated movements. This theory has already strained people's imagination. According to it, objects become heavier and time elapses more slowly the faster they move, space, more precisely space-time, is curved, and objects do not move in a straight line without any force applied, but their movements follow curved paths. These unusual behaviors of physical objects cannot be observed in our everyday life on earth, which is why they defy common sense and appear so strange. But the theory of relativity has been convincingly proven, when observing large masses and high speeds in space and using special equipment on earth. As unusual as its predictions may seem, they have been confirmed experimentally time and time again. The statements of the theory of relativity are therefore considered to be well founded, and it is *the* theory when it comes to describing large objects.

At the other end of the length scale, we have the world of the tiniest particles. The theory of relativity seems unusual, but the quantum theory seems bizarre. According to this, the quanta, the smallest objects that make up our world, are both waves and particles – at the same time. In addition, it is not possible to specify the path of a particle or its properties, but these are only known with a certain probability before the measurement is performed. It is only after the measurement that we actually know the properties of the quantum. Quantum theory is therefore referred to as "new" physics, where we no longer have clear determinism. The opposite is true in classical physics, in which the movement of each particle can be described precisely. The theory of relativity is also part of classical physics, no matter how unusual its predictions may be, because it also allows the trajectory of a particle to be determined precisely.

The theory of relativity as well as the quantum theory are very successful in their fields. Thanks to relativity, we can calculate times of moving objects so precisely that GPS satellites can determine our position on Earth with great accuracy. Thanks to relativity, we can write down the history of the universe and the stars. Thanks to quantum theory, we are able to make modern computers and lasers. Our modern technology would not be possible without quantum theory. But despite these successes, physicists are not satisfied. This is because the theory of relativity and quantum theory are not compatible with each other. The theory of relativity does not allow us to describe the quantum world, and quantum theory fails completely in describing large objects. Physicists are therefore looking for a theory that describes both worlds at the same time, a "theory of everything". This would also allow to describe extreme objects like black holes or the big bang, for which not only the theory of relativity is needed, but also the quantum theory, because these objects are not only extremely massive, and thus fall into the field of relativity, but also extremely small, and thus are objects of the quantum world.

The basic ideas of the theory of relativity and the quantum theory are known since the beginning of the 20th century. Since about 100 years physicists are searching for a unification of these two ideas, but this search has not been successful so far. Both worlds seem to be much too different as that a common denominator could be recognized.

However, the view of physics is obscured by the fact that the quantum world is considered solely through the eyes of the so-called Copenhagen interpretation. Many paradoxes of the quantum world, which make it so difficult to establish a connection to relativity, are in fact not paradoxes of the quantum world, but paradoxes of the Copenhagen interpretation. If we want to connect the quantum theory with the theory of relativity, it must become clear what are really properties of the quantum world – and what are only properties of a special interpretation of the quantum world.

This book presents some paradoxes of the quantum world and discusses whether these are really paradoxes of the quantum world or rather paradoxes of the Copenhagen interpretation. By separating the wheat from the chaff, we can focus on the actual characteristics of the quantum world and will thus obtain some properties that are undoubtedly associated with the quantum world, while others that we have taken for granted as belonging to the quantum world are in fact only characteristics of the Copenhagen interpretation. Once we have done this, it seems possible to identify a common denominator that can connect quantum theory with the theory of relativity.

2. Quantum mechanics – a short introduction

By the end of the 19th century, physics had essentially explored the field of mechanics, electromagnetic phenomena were comprehensively described by Maxwell's equations, and thermodynamics dealt with otherwise elusive properties of nature such as "energy" and "heat". Many a physicist thought that physics had explained the whole world in principle, and that there was nothing left for physicists to do except for the more detailed description of some minor details. Therefore, in the second half of the 19th century, a young man was advised not to study physics, because there would be nothing left to discover. The young man who was given this advice was Max Planck. And it was this man who would be standing at the beginning of a revolution that would shake physics.

It was therefore a stroke of luck for physics that Planck did not take this advice and studied physics. He chose thermodynamics as his main focus, as this subject had received a new impetus at the end of the 19th century by Ludwig Boltzmann. Boltzmann had described the macroscopic magnitude of entropy in terms of probabilities, more precisely: in terms of the number of microstates that the particles in a system could assume (Boltzmann 1877), thus connecting the micro with the macro world. This interpretation of thermodynamics was not well received by most other physicists. On the one hand, many physicists did not like the fact that it was no longer possible to describe exactly how a system behaved, but had to work with statistical statements, and on the other hand, they were not convinced that these small particles, these atoms, actually existed. They could be used well for calculations, and in chemistry they explained the composition of molecules, as the chemist John Dalton had shown. But did they really exist?

By and by, however, everybody accepted that atoms actually exist. This was not the case, however, because scientist could be convinced, but as Max Planck put it so beautifully: "A new scientific truth does not usually assert itself in such a way that its opponents are convinced and declare themselves to be enlightened, but rather by the fact that their adversaries are gradually dying out and that the rising generation is familiar with the truth from the start."

At the end of the 19th century, Planck turned to the radiation of a black body. A black body is an ideal absorber, because it absorbs all radiation. Since such a body cannot heat up endlessly, it must therefore also be an ideal radiator, emitting the optimal spectrum of light at a given temperature. The simplest way to make a black body that emits the optimal spectrum is to take a cavity with a small hole from which light can escape, but through which hardly any light enters the cavity. The cavity is heated to a certain temperature, and the light that escapes corresponds to the light of a black body and can be measured through the hole.

It was quickly established that the light distribution has a maximum, which changes with temperature: the higher the temperature, the smaller the wavelength of the maximum. At low temperatures, the maximum is in the red area; if the temperature is increased, it moves into the yellow, green and finally blue area of the spectrum.

Of course, scientists wanted to explain the spectrum of the black body with the means of the physics of that time. The black body emitted light, thus electromagnetic waves. These are created in the cavity by the supplied heat. At the walls of the cavity the wave must have a node, otherwise there are not further limitations. If you want to determine the energy density of the electromagnetic radiation, you only have to sum up all the waves which can form in the cavity. But this number is not limited. The wavelengths become smaller and smaller, which means their energy increases, but there is no smallest wavelength. Thus, the energy density increases with smaller wavelength. Therefore, at the smallest wavelengths, the energy density should become infinite. This is of course nonsense. A hollow body which is heated with finite energy cannot emit infinitely high radiation energy. The theoretical prediction that this should happen was later called the "UV catastrophe". So, simply adding up the waves that can theoretically form in a cavity doesn't work. It thus remained a mystery how the spectrum of a black body could be explained theoretically.

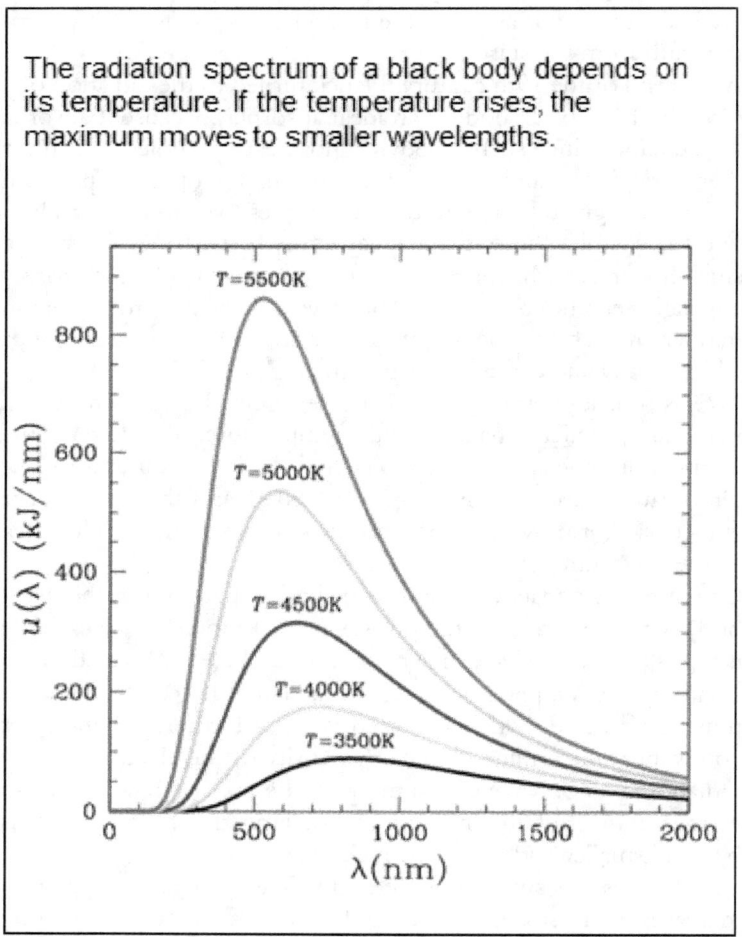

The radiation spectrum of a black body depends on its temperature. If the temperature rises, the maximum moves to smaller wavelengths.

There can be any number of waves with ever smaller wavelengths in a cavity. The total energy calculated in this way can be as large as you like, resulting in a UV catastrophe.

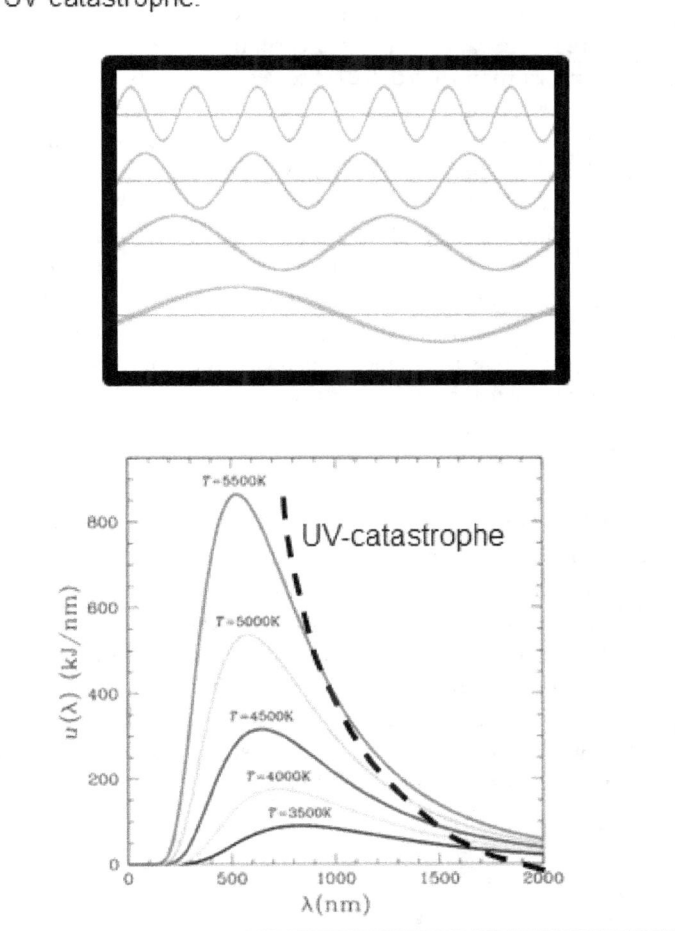

Planck found a solution by choosing an unusual approach: He did not sum up every wave, but he only took those waves whose energy was a multiple of a constant, later called Planck's quantum of action, as if the energy of the waves did not change continuously, but in small portions. With this approach, Planck obtained a formula that described the spectrum and the observed variation of energy density as a function of the wavelength.

The starting point for Planck's considerations was that the temperature in the cavity leads to a certain average thermal energy. Planck now assumed that there were resonators in the cavity that oscillated at a specific frequency v. According to Planck, the energy of the resonator was $E=hv$, where the variable h originally stood for "help", as it helped Planck to solve the riddle. If the energy of a resonator was below the mean thermal energy, then it was definitely excited. For small energies and large wavelengths, you could simply add up the vibration modes, as in the classical case. However, if the energy of a resonator was above the mean thermal energy, then the resonator was only excited with a certain probability, which was smaller the more the energy required to excite the resonator was above the mean thermal energy. The greater the energy, the smaller the number of oscillations, since the probability of excitation decreases. In contrast to the classical summation, this assumption does not lead to a UV catastrophe, but the spectral energy density decreases again at high frequencies, i.e. small wavelengths. In this way, Planck obtained a formula that correctly described the spectrum of the black body.

At the end of 1900 Planck presented the result of his calculation (Planck 1900a, 1900b). Today, this is considered to be the birth date of quantum mechanics. At first, Planck thought that his approach was only a mathematical trick; he did not believe that the energy of waves actually exists only in small portions. This step was taken by Albert Einstein in 1905.

According to Planck, the radiation is quantized and can only assume certain values. If the energy required to excite the radiation is greater than the mean thermal energy of the black body, then the state cannot be occupied. For higher frequencies, the occupation probability decreases again. N is the number of particles.

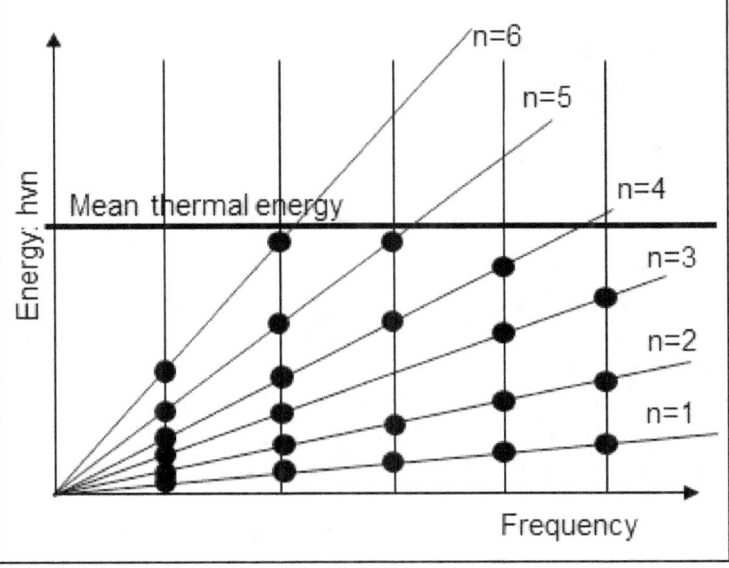

With the assumption that light occurs in quantized portions, Planck obtained a formula for the spectral energy density of the radiation of a black body that described the observed behavior:

$$U(v, T)dV = \frac{8\pi h v^3}{c^3} \frac{1}{e^{hv/kT} - 1} dv$$

With:
h Planck's constant,
k the Boltzmann constant,
v the frequency,
T the temperature and
c the speed of light

At the beginning of the 20th century, Einstein looked at the photoelectric effect, which had been discovered at the end of the 19th century (Einstein 1905c). Researchers such as Henry Becquerel and Heinrich Hertz had found that electromagnetic radiation, such as light, that hit a metal plate changed it electrically. Philipp Lenard examined this effect in more detail in 1899. He found that electrons leave the metal when the metal is exposed to light.

At first side, this effect has nothing unusual. It has been known for a number of years that electrons are part of atoms, even if it was not yet known what atoms really looked like. Light is nothing more than radiant energy, so it was easy to imagine that irradiating electrons would give them enough energy to overcome the attraction to the positive atoms and leave the metals. You only had to irradiate a metal surface long enough so that the electrons could absorb enough energy from the electromagnetic waves to leave the metal; light with a longer wavelength, which has lower energy, would just take a little longer than light with a short wavelength. However, this was not observed.

The experiments showed that there was a cutoff wavelength for each metal. If the light had a shorter wavelength, then free electrons were created; if it had a longer wavelength, then you could shine on the surface for hours and still not observe any free electrons. This could not be explained in the understanding of classical physics, where light consists of waves that continuously provide energy.

In the photoelectric effect, light knocks electrons out of a metal. However, this only works if the light has a certain minimum energy. If light were a wave, the energy would accumulate over time and one would only have to wait long enough to get free electrons. Since this is not the case, Einstein concluded that light consists of particles that collide with an electron and transfer their energy in the process. If the energy is too small, the electron cannot overcome the attraction of the metal.

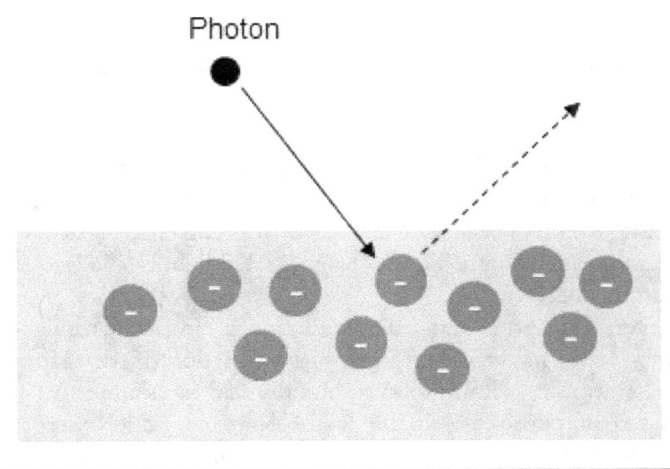

Photon

Einstein assumed that light actually consisted of energy quanta, which were therefore real, even if Planck had only thought of them as a mathematical hypothesis. Later, these light quanta were to be given the name "photons". If such a quantum of energy hits an electron, then it can transfer its energy to the electron. If the energy is large enough, the electron can overcome the attraction of the atom and it escapes from the metal. If the energy of the photon is too small, then this energy is not sufficient and the electron remains trapped in the atom. With the assumption that light consists of energy quanta, Einstein was able to explain quite simply why there was a cutoff frequency of light, with electrons escaping for higher frequencies immediately and remaining trapped forever for lower frequencies. Einstein's explanation of the photoelectric effect was therefore quickly accepted.

Further evidence that light consists of particles was provided by Arthur Compton, who conducted some experiments in 1922 in which he shot high-energy X-rays at graphite and examined the scattered light. He made the observation that the wavelength of the scattered radiation was longer than that of the incident radiation. The X-rays had therefore lost energy when scattering on the graphite.

This observation could not be explained with the assumption that light was an electromagnetic wave, either. Because then the wave would excite an electron, which would then emit a wave with the same wavelength. It was not clear why the wavelength should change.

Even more inexplicable was a second observation: The wavelength did not change randomly, but it depended on the scattering angle. The larger the scattering angle, the greater the decrease in wavelength. The assumption that light is a wave cannot explain this behavior.

However, Compton was able to derive a simple formula that described the change in wavelength as a function of the scattering angle (Compton 1923). For this, he only had to assume that light consisted of particles colliding with electrons, which were also particles. With the help of the conservation of momentum,

Compton could then describe the change of the wavelength. Thus, Compton presented a further experimental proof that light consists of particles.

If light collides with a particle, in this case an electron, the wavelength of the radiation changes depending on the scattering angle. This can only be explained if light consists of particles that lose energy in a collision process. The following applies:

$$\Delta\lambda = \frac{h}{m_e c}(1 - \cos\varphi)$$

With:
Δλ the change in wavelength,
h Planck's constant,
c the speed of light,
m_e the mass of the electron and
φ is the scattering angle

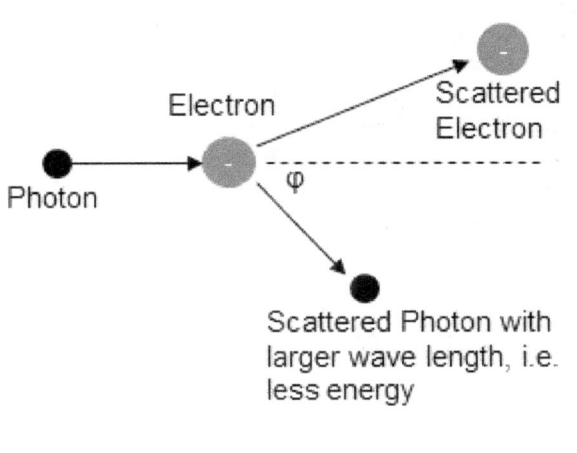

However, with the assumption that light consists of photons, Einstein also took a first step towards a problem that would later worry quantum mechanics extensively. Before Einstein, there was no question that light was made up of waves. Maxwell's equations of electrodynamics describe electromagnetic waves, which include light, and Thomas Young's early 19th-century double-slit experiment clearly showed that light must consist of waves, since only they could explain the observed interference pattern. Before that, there had been two theories: Isaac Newton believed that light was made up of small particles, and Christian Huygens believed that light was made up of waves. Initially, Newton, the founder of mechanics and co-founder of calculus, prevailed because of his greater authority. But then, in 1802, Thomas Young performed his famous double-slit experiment – and after that no one seriously considered any longer that light could be made of particles.

In this experiment, Young directed the light from a narrow light source (narrow so that the light would be in phase; because lasers did not exist at that time) onto a cardboard in which he had cut two slits. Behind the cardboard was a screen. If he covered one slit, then the part of the screen behind the open slit shone brightly, if he covered the other slit, then the wall behind the first slit shone brightly.

Then he left both slits open. Should the light consist of particles, then the particles would either go through the first or the second slit, and one would expect two bright stripes on the screen. If light is made up of waves, then the waves should overlap (which is why the light had to be in phase and the light source narrow), and one would expect a pattern of light and dark fringes depending on whether the waves overlap constructively (wave crest on wave crest) or destructively (wave crest on wave trough). Young observed an interference pattern of light and dark fringes that caused the brightness maximum to lie between the two slits, making it clear that light must consist of waves.

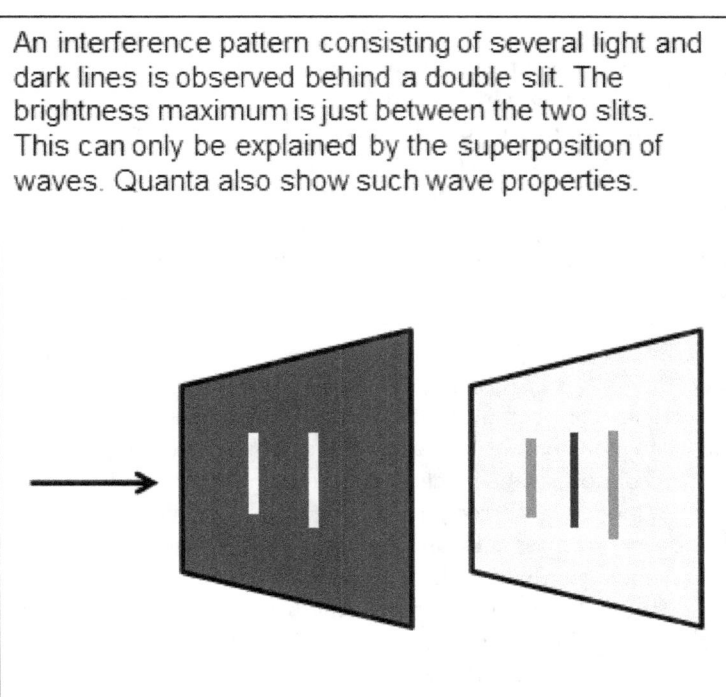

An interference pattern consisting of several light and dark lines is observed behind a double slit. The brightness maximum is just between the two slits. This can only be explained by the superposition of waves. Quanta also show such wave properties.

And now Einstein came and said that it consisted of particles. But it also consisted of waves without any doubt. What, therefore, was correct? How should you imagine light quanta which could be both wave and particle?

Even though Einstein's explanation of the photoelectric effect raised the question of how quanta could exhibit properties that could sometimes be interpreted as waves and sometimes as particles, this question was not further pursued at first. It became an urgent problem only as physicists wanted to understand the structure of atoms.

At that time, atoms were understood to be the indivisible components of matter; after all, the Greek word "atomos" means indivisible. That this could not be true was announced in 1897 by Joseph John Thomson when he discovered the electron as a component of the atom. Since the electron is negatively charged, but the atom is neutral, there had to be positive building blocks in the atom, the protons. The atom thus could not be the smallest building block of matter. However, it was unclear how protons and electrons are arranged in the atom.

Thomson himself developed an atomic model which was referred to as the "plum pudding" or "raisin cake" model. According to this model, the atom is a positive mass in which the electrons are distributed, comparable to a cake in which the raisins are arranged. Starting in 1909, Ernest Rutherford carried out scattering experiments on a gold foil. In doing so, he shot alpha particles onto the gold foil. Alpha particles are positively charged particles that result from certain types of radioactive decay. As we know today, an alpha particle consists of two protons and two neutrons, it corresponds to the nucleus of a helium atom. If Thomson's atomic model is correct, then the atoms should be a largely homogeneous mass, allowing the alpha particles to fly through the foil essentially unhindered.

However, Rutherford observed something different: Most of the particles actually flew through the gold foil unhindered, but a few alpha particles were scattered back, as if they had hit a hard obstacle from which they bounced back. Rutherford was able to estimate that this hard core had a diameter of about one-thousandth of the diameter of the atom. When he repeated the experiment with other materials, he also found that the positive electric charge of the nucleus corresponded to the atomic number of the element of the foil. He then developed a "planetary model" of the atom: the positive charges are located in the nucleus, and the electrons move around this nucleus, just as the planets move around the sun. The atom thus consists essentially of nothing, since the tiny electrons almost disappear in the large outer region of the atom (Rutherford 1911).

When Rutherford shot alpha particles at a gold foil in his scattering experiment, most of the particles flew through almost unhindered, but some were also reflected back as if they had encountered a hard obstacle.

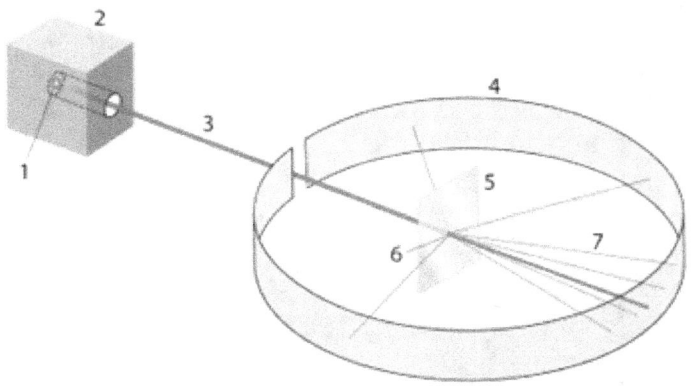

With:
1: radioactive radium,
2: lead sheath for shielding,
3: alpha particle beam,
4: fluorescent screen
5: gold foil
6: point where the rays hit the foil,
7: Particle beam after penetrating the gold foil, a few are deflected back

Rutherford concluded from his scattering experiment that the atom consists of a small, positive nucleus around which electrons move, with the charge on the nucleus equaling the total charge on the electrons. The atom should thus resemble the structure of our solar system, in which the planets revolve around the sun.

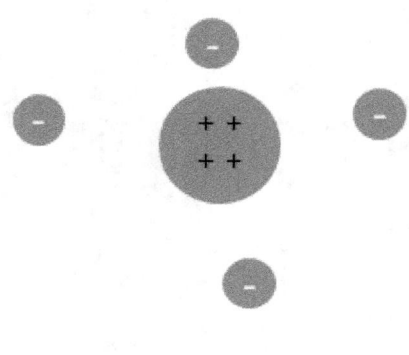

While this model fitted the observations, there was a major problem: the electrons are negatively charged. When they move around the atomic nucleus like planets around the sun, they constantly change their direction (since they don't fly in a straight line, but on a circular path), so they are accelerated. However, an accelerated electric charge radiates electromagnetic waves, causing it to lose energy. The electrons would therefore be losing energy, and their orbits would bring them closer and closer to the nucleus until they fell into it. If Rutherford's atomic model were correct, atoms would cease to exist after a few fractions of a second. However, atoms apparently exist much longer, so Rutherford's model of the atom could not be correct, although it was consistent with the observation that there is a hard nucleus at the center of the atom.

Niels Bohr proposed a solution that took up Planck's and Einstein's idea of quantized energy (Bohr 1913). He assumed that there were certain orbits around the nucleus where the electrons did not lose energy. Then he claimed that electrons absorb and release energy as they move between these orbits. Thus, the radiation emitted by an atom is not determined by the orbital frequency of the electrons around the nucleus, as one might think coming from classical electrodynamics, but by jumps between different orbits. In addition, he made the claim that the angular momentum of the electrons is also quantized, i.e. it can only assume values which are a multiple of Planck's quantum of action. With these assumptions he obtained a formula for the energy of the electron as a function of its orbit around the nucleus. The difference between two orbits provided energies that corresponded exactly to the spectral lines of hydrogen. This formula for the spectral lines had already been empirically derived by Johannes Rydberg in 1888, but he had not been able to explain it. Bohr could explain with his simple model how the spectral lines of hydrogen were created.

However, Bohr's simple model failed for other atoms. In addition, it could not explain the fine structure of the spectral lines. However, Arnold Sommerfeld succeeded in doing this in 1916 (Sommerfeld 1916a, b). He extended the model from circular to elliptical orbits. In addition to the main quantum number n, he needed two other quantum numbers for this, the secondary quantum number l and the magnetic quantum number m_l. Later it was recognized that the electrons also have a spin, so that electrons in an atom are described by a total of four quantum numbers.

With these additions, all the intricacies of the spectra of hydrogen could be described, but these formulas still did not work for other atoms. It also remained completely unclear why there should be orbits at all on which the electrons did not radiate any energy. Louis de Broglie provided a possible answer in his 1924 doctoral thesis.

Niels Bohr assumed that the electrons can only move around the nucleus in certain orbits on which they do not lose any energy. If they move from a higher to a lower orbit, they emit energy in the form of radiation – and we perceive this energy difference as the spectral line of the element.

Bohr received the following formula for the energy of the quantized electron orbits:

$$E_n = -\left(\frac{e^2}{4\pi\epsilon_0}\right)^2 \frac{m}{2\hbar^2}\frac{1}{n^2}$$

n is the main quantum number.

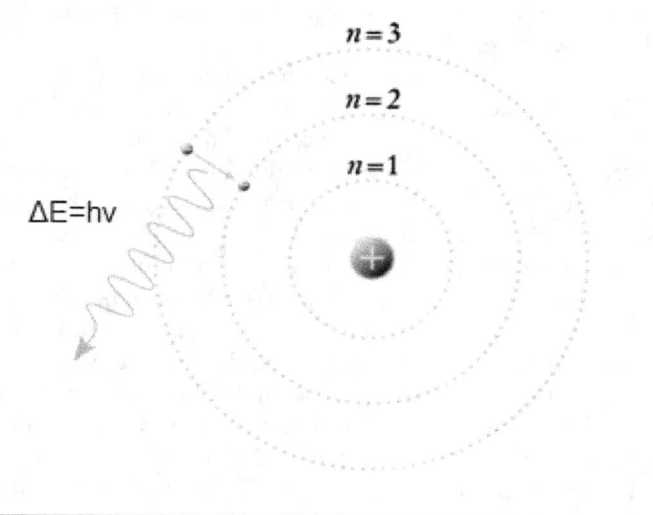

De Broglie had noticed some similarities between the dynamics of particles and the dynamics of waves, and he therefore made the claim that particles with a certain momentum could be assigned a certain wavelength (de Broglie 1925). The momentum is inversely proportional to the wavelength $(p=h/\lambda)$, i.e. the smaller the wavelength, the larger the momentum.

De Broglie's investigations are usually cited in a shortened version which says that matter particles can also show wave properties. In fact, de Broglie never said this. He laid the foundation for an interpretation of quantum mechanics that is fundamentally different from the Copenhagen interpretation, as we will see later. However, he published his results only in French, which might explain why his results were mainly known in a shortened version. But if particles like the electrons show wave properties, then one should be able to observe interference patterns as with light. However, the wavelength of the electrons is very small. In order to show wave properties such as interference, a grating with correspondingly small distances was needed, which was technically impossible to produce at the beginning of the 20th century. But nature supplies these grids free of charge, namely in the form of crystals. In 1927, Clinton Davisson and Lester Germer fired electrons at a nickel crystal and were able to observe the interference pattern typical which is typical for waves. Matter also shows wave properties (Davisson Germer 1927).

So now we have the fact that particles can also show wave properties, just as, according to Einstein, light waves can also show particle properties. For light waves a wave equation is known. If particles can also show wave properties, then there should also be a wave equation for matter particles. Erwin Schrödinger followed this idea and in 1926 presented his famous equation (Schrödinger 1926a, 1926b). This equation allows to calculate wave functions and the energies of electrons in the atom. With the assumption that electrons are waves, it was also possible to explain why there were only certain states and energies in which electrons existed in an atom: Only here does the wave function in the atom superimpose itself constructively, i.e. it does not cancel itself out. These are the only possibilities for stable orbits; electrons cannot take other orbits.

But what is the wave function actually? It is not a wave like the light wave, which propagates in space and directly describes the behavior of the electromagnetic radiation; because the wave function is a complex function.

The physical world is described with real numbers like 3 or 4.125. Complex numbers include the root of -1, which has no solution in the area of real numbers. This root is usually abbreviated with the letter i. A complex number is written for example as 3+4i. Complex numbers are often used in physics because some calculations are easier with them than with real numbers. However, in order to get physically meaningful values, one must translate the complex functions back into real functions. The complex functions do not describe physics. But the wave function is a complex function. Which kind of physics does it describe?

Max Born found a way out of this dilemma: In his opinion, one has to calculate the square of the absolute value of the wave function (this eliminates the complex numbers) which then provides a distribution for the probability that a particle is in a certain state. According to this interpretation of quantum mechanics, one cannot calculate exactly defined states (where a quantum is located), but only how probable a certain state is (with what probability a quantum is located at a certain location).

With this procedure, one obtains from the Schrödinger equation probability densities for the electrons, the so-called orbitals. The orbitals are described by three quantum numbers. The principal quantum number n determines the energy of a shell, which can contain several orbitals. The orbital angular momentum quantum number l determines the shape and number of orbitals. The rule is that $l<n$. If $l=0$ then the orbital is spherical, if $l=1$ then there are three dumbbell shaped orbitals. This quantum number is also denoted by letters. The case of $l=0$ is called s-state, for higher values of l the names are p, d and f. The magnetic quantum number m_l determines the orientation of the orbitals. It can take the values -1, ..., 0, ...1. If there are three orbitals, then they align along the three spatial directions. Last but not least there is the spin quantum number m_s, which can assume the values ½ and –½. According to the Pauli rule, two electrons in an atom cannot have the same four quantum numbers. With this knowledge, starting with the lightest atom, hydrogen, it is possible to fill the shells and orbitals for the

different quantum numbers and determine the structure of the electron shell.

The Schrödinger equation is:

$$i\hbar \frac{\partial}{\partial t} \Psi(t) = H\Psi(t)$$

with:
$\hbar = h/2\pi$,
H the Hamilton operator describing the energy of the system, and
$\Psi(t)$ of the wave function.

For the hydrogen atom one obtains solutions of the Schrödinger equation of the form:

$$\Psi_{nlm}(r, \vartheta, \varphi) = R_{nl}(r) Y_{lm_l}(\vartheta, \varphi)$$

The radial function R depends only on the radius and the two quantum numbers n and l, and the angular function Y on the angles and the quantum numbers l and m_l. The associated probabilities are spherical or dumbbell-shaped:

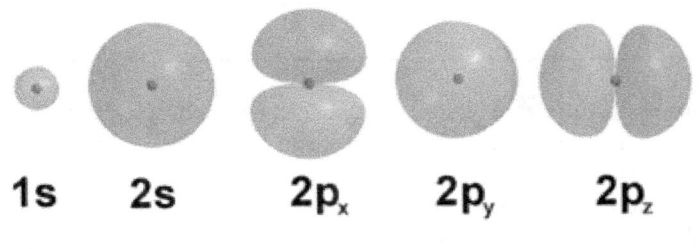

1s   **2s**   **2p**ₓ   **2p**_y   **2p**_z

The quantum numbers describe how the shells and orbitals of atoms are filled. It should be remembered that there is still the spin quantum number m_s which can assume the values $+\frac{1}{2}$ and $-\frac{1}{2}$, and that the Pauli rule states that the electrons in an atom must not have the same quantum numbers.

The periodic table shows which shells and orbitals are filled. For n=1 there is only the s orbital. For higher shells (=periods), first the s orbitals of groups 1 and 2 are filled, then the p orbitals of groups 13 – 17. The d orbitals are filled for the subgroup elements of groups 3 – 12. The f orbitals are filled for the lanthanides and actinides.

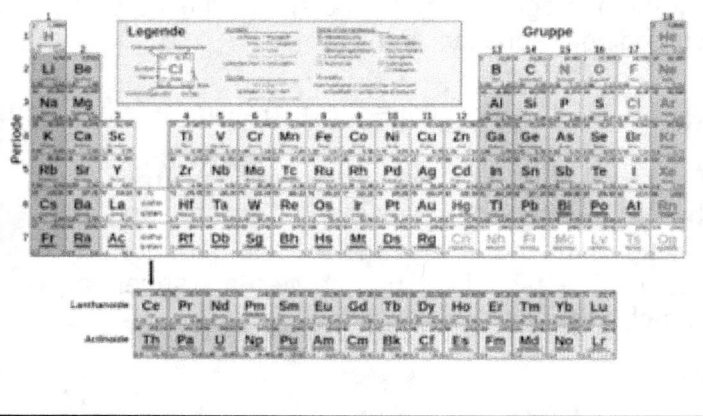

This procedure to obtain orbitals from the Schrödinger equation describes very well the structure of the elements, but is somehow unsatisfactory: You start with the Schrödinger equation, which is supposed to describe the physics of a quantum system, but the solution is first of all a physically nonsensical wave function. Only if you square it, you get a physically meaningful solution – which, however, also provides only probabilities about the behavior of the particles and no unique value, as one is used to from classical physics.

This fact, that you can get only probabilities in quantum mechanics, is often justified with the Heisenberg uncertainty principle.

Already in 1925, Werner Heisenberg had developed a calculation method to determine the energies of the system (Heisenberg 1925), which is based on the mathematical formalism of matrices. However, since this formalism is much more abstract than the Schrödinger equation and also delivers the same results as the latter, we will limit our discussion to the Schrödinger equation.

In 1927, Heisenberg presented his uncertainty principle in another paper (Heisenberg 1927). According to this, it is not possible to measure precisely and at the same time the position and the momentum of a particle, or the energy and time of a process, but the uncertainties are in the order of magnitude of Planck's constant. Heisenberg initially illustrated this uncertainty principle as a measurement problem: If we look at a car, we can do this because photons are reflected by the car and end up in our eyes. When reflected, the photons transfer an impulse to the car. However, the mass of the car is so large and the momentum of the photons so small that this has no influence on the measurement.

This is different when we observe a quantum particle like an electron. If we observe it with the help of a photon, then we get a knowledge about where it is. However, the photon transfers its momentum to the electron, which changes its momentum significantly. Location and momentum cannot be measured at the same time.

Later, the uncertainty in the quantum world was not understood as a measurement problem, but as a fundamental property of the quantum world. It is not possible to measure location and momentum simultaneously in principle. The measurement of location and momentum are mutually exclusive, these two variables are "complementary" as Bohr called it. The reason is that a quantum reveals itself as a wave or a particle. As a particle we know its location, as a wave its momentum. But both at the same time cannot be measured. In the measurement we receive information about either the wave or the particle property. But what, then, is a quantum?

Heisenberg's uncertainty principle states that the uncertainty in the measurement of position q and momentum p is of the order of Planck's quantum of action.

$$\Delta p \cdot \Delta q \approx h$$

The same applies to the energy E and time t of a process.

$$\Delta E \cdot \Delta t \approx h$$

The space-momentum-uncertainty can be illustrated with diffraction at the slit: The smaller the slit, the further apart are the diffraction maxima, the distances between which are determined by the momentum.

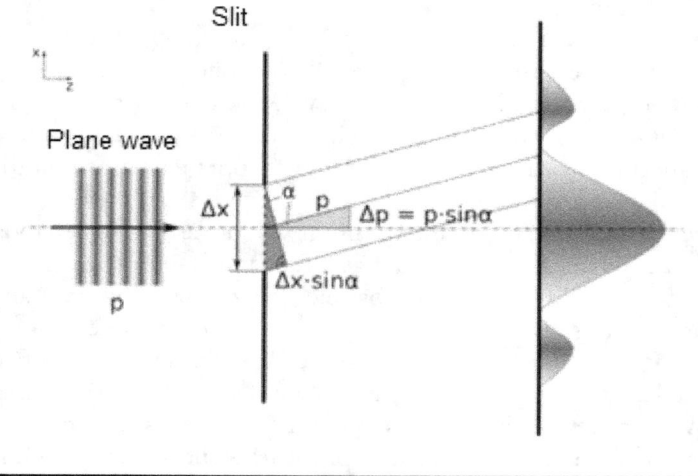

Bohr and his colleagues like Heisenberg, who developed the so-called Copenhagen interpretation of quantum mechanics, were convinced that a quantum is undetermined before a measurement. It is anything. It's the measurement that tells us what it is, like a dark room could contain anything – a chair, a closet, a bed – and only when we turn on the light, when we perform the measurement, we will find out that there is a chair in the room. The other possibilities, which were just as real until the measurement and also existed with their respective probabilities, disappear at this moment, like picking exactly one card from a deck of cards and discarding the others. All you can tell before taking a measurement is the probability that there is a chair, a bed, or a closet in the room. According to the Copenhagen interpretation, the quantum world is not deterministic. You cannot say what is available before a measurement. The result is only created during the measurement.

Not everyone liked this interpretation of quantum mechanics. Especially Schrödinger and Einstein had difficulties with the idea that a system should be undetermined until you measure it. After all, the moon moves around the earth even if you don't look at it right now – at least this can be assumed with high probability. According to Einstein and Schrödinger, the fact that we can only describe the quantum world with probabilities is due to the fact that we cannot measure the quanta exactly – precisely because they are so small. We have the same problem as in thermodynamics, where you can describe the behavior of the systems also only statistically. Nevertheless, one assumes that the particles of the matter are present all the time. When we measure a chair in the room, then it should have been there all the time – and did not just materialize in the measurement.

Schrödinger's wave function (more precisely: its square) describes the probabilities for the presence of an electron. Classically, as Schrödinger and Einstein did, you would expect the electron to exist somewhere as a particle, we just can't say exactly where, so we have to describe it in terms of probabilities. According to the Copenhagen interpretation, the electron exists as a probability wave before the measurement, and thus everywhere, and only becomes a particle during the measurement. Einstein did not like the idea that the probability in the quantum world is not a result of our ignorance, but a fundamental phenomenon that makes a precise knowledge of the quantum world impossible, which is why he said: "God does not play dice."

But by the end of the 1920s, the majority of physicists had come around to Bohr's views and adopted the Copenhagen interpretation. Physicists no longer try to describe the behavior of a system in all details, but only calculate the results, which also became known as "Shut up and calculate!"

Unsatisfying, however, are the paradoxes that arise from this understanding of the quantum world. Some of the most famous ones were raised by Schrödinger and Einstein and shall be discussed in the following section.

3. Paradoxes of the quantum world

3.1. The collapse of the wave function

According to the standard interpretation of quantum mechanics, the quantum is undetermined before the measurement. It is not only unknown because we cannot measure its parameters exactly, just as we cannot measure the parameters of all particles in a liter of gas exactly, but it is undetermined because the quantum does not exist as a particle before the measurement. It exists before the measurement as a superposition of all possibilities, as if the object in the dark room was in equal parts a chair, a closet and a bed. We can only know what it is after we have done our measurement. During the measurement one possibility manifests itself, the others disappear.

This is called the collapse of the wave function. Before the measurement the quantum exists as superposition of all possibilities, afterwards there is only one. This superposition is part of the Schrödinger equation as it is linear: If you have two solutions of the Schrödinger equation, then the sum of these two solutions is also a solution, you can combine it linearly from the two solutions.

However, this understanding of the measurement process gives rise to a problem that Einstein in particular had to struggle with: Before the measurement, a particle can theoretically be everywhere: directly in front of the measuring device, but also on Jupiter. The quantum's wave function is very, very small on Jupiter, but it is not zero, so the square of the absolute value, the probability that it is on Jupiter, isn't zero, either. Instead of measuring it on Earth, you could also measure it on Jupiter. The wave function and thus the quantum is spread out between Earth and Jupiter, even throughout the universe.

Then you perform the measurement where you measure the particle on Earth, and at that very moment the wave function on Jupiter also ceases to exist. The quantum practically contracts to the point where we measured it, and it does so instantaneously.

The wave function describes the behavior of a quantum. Since the wave function of a free quantum is not limited, the particle can be found both in the laboratory and on Jupiter, albeit with a lower probability. When we perform a measurement, it will be measured in the laboratory and the wave function on Jupiter disappears. It collapses – instantaneously and thus faster than the speed of light.

Analogously, a spreading wave (here a sea surface) would collapse instantaneously to a point.

This is also the point that caused Einstein so many headaches: The theory of relativity is based on the idea that nothing can propagate faster than the speed of light. An instantaneous change over any distance is therefore not possible. Such a change is also called non-local, while changes that are slower than the limits of relativity are called local changes.

According to Einstein, it simply cannot be that the wave function contracts instantaneously to a point. Bohr and the other physicists who supported the Copenhagen interpretation had no problem with this. In their view, a quantum is not defined before measurement, so there is no need to worry about what a quantum looked like before. And since it is not defined before the measurement, there is no point in worrying about whether somehow the theory of relativity might be violated.

But if you cannot say what a system looks or behaves like before the measurement, then it is of course also difficult to unite this theory (of quantum mechanics) with another theory (like the theory of relativity).

Not only Einstein had problems with the collapse of the wave function and the fact that a system can take all states at the same time before the measurement. Schrödinger also found this idea absurd and developed an example to illustrate the absurdity of this idea.

3.2. Schrödinger's cat

This example is the famous Schrödinger's cat. He first described it in an article he published in 1935 (Schrödinger 1935b, p. 812):

> One can also construct completely burlesque cases. A cat is locked in a steel chamber, along with the following infernal machine (which must be protected against direct access by the cat): In a Geiger counter tube there is a tiny amount of radioactive substance, so little that in the course of an hour maybe one of the atoms decays, but just as likely none; if this happens, the counter tube responds and actuates a hammer via a relay, which smashes a flask of hydrocyanic acid. If you have left this whole system to itself for an hour, you will tell yourself that the cat is still alive if no atom has decayed in the meantime. The first nuclear decay would have poisoned them. The wave function of the whole system would express this as having

the living cat and the dead cat mixed or smeared in equal parts.

Quantum mechanics would therefore assume that the cat exists in the two states dead and alive at the same time. Only when the chamber is opened does the wave function collapse and one receives the information as to whether the cat is dead or alive.

If the Copenhagen interpretation were true in the classical world, one would have to assume that a cat placed in a box with a randomly controlled death machine is both dead *and* alive unless you open the box and check.

For the proponents of the Copenhagen interpretation, Schrödinger's cat is a proof of how bizarre the quantum world is. For Schrödinger, this thought experiment was an indication that the Copenhagen interpretation cannot make sense.

The idea that a cat can be dead and alive at the same time is of course absurd. And Schrödinger was convinced that this is not only absurd for a macroscopic object like the cat, but must be absurd for the quantum world as well. An object cannot assume several states at the same time, even if the equation he developed allowed it. But this was just the question about the interpretation of the equation: Does the fact that the equation also allows a superposition of states actually mean that these superpositions exist – or is this only an expression of the fact that one does not know exactly in which state a quantum is because you cannot measure its properties precisely enough?

Schrödinger and Einstein advocated this interpretation. In their opinion, the state of a quantum had to be precisely defined, even if we are not able to measure it precisely with our crude measuring instruments. However, Bohr, Heisenberg and other proponents of the Copenhagen interpretation assumed that the state of a quantum is undefined before the measurement. A quantum can actually be in all possible states at the same time, even if this may seem absurd in the macroscopic world, as the example of Schrödinger's cat shows.

3.3. Wigner's friend

In 1961, Eugene Wigner took Schrödinger's thought experiment to the extreme. Wigner imagined that he was working with a friend in a laboratory where a quantum experiment was being performed (Wigner 1961). The friend was in the lab observing the measurement, Wigner was outside the lab and could only learn from his friend how the measurement had turned out.

As long as the friend did not perform the measurement, the quantum system is in a state of superposition, as is commonly assumed. Let us suppose that it can assume two states, like Schrödinger's cat. Then, before the measurement, the system is in state 0 and state 1 at the same time. Only when Wigner's friend makes a measurement does it assume a concrete value. But strictly speaking only for Wigner's friend – and here it becomes really bizarre.

Because Wigner's friend now knows what state the system is in, but Wigner doesn't know it yet. For him there is still a superposition, because he doesn't know what his friend has measured. He only knows that according to the laws of quantum mechanics the system should be a superposition of "system is in state 0 / friend measured 0" and "system is in state 1 / friend measured 1". Only when he has asked his friend does Wigner know whether the system is in status 0 or 1.

Thus, you have the bizarre situation that the quantum experiment has delivered a clear result for Wigner's friend, but is still undetermined for Wigner. The wave function has collapsed for Wigner's friend, but not yet for Wigner. The wave function exists and has collapsed at the same time. How can that be?

If Wigner's friend has carried out an experiment, then the wave function has already collapsed for him. Wigner, who is waiting outside the laboratory, does not yet know the result – for him the wave function still exists. The wave function can therefore exist at the same time – and no longer exist.

Some physicist tried to explain this paradox by saying that the superposition is not destroyed by the measurement but only by the perception of a consciousness. As long as a consciousness has not perceived a measurement, the superposition can still exist. Therefore, the wave function had already collapsed for Wigner's friend, since his consciousness had already perceived the measurement, whereas the superposition still existed for Wigner, since his consciousness did not know the result of the measurement.

It is said that the famous physicist Richard Feynman has asked whether the consciousness of a cleaning lady is enough for the collapse of the wave function or whether it has to be a quantum physicist.

3.4. The double slit

The quantum world is bizarre. So far, we've only scratched the surface. It gets even more bizarre if you take a closer look at the double-slit experiment. Incidentally, Richard Feynman was of the opinion that the double-slit experiment is *the* ultimate puzzle of quantum mechanics. If you have solved this, according to Feynman, then you have actually understood the quantum world.

The double-slit experiment is more than 200 years old and served to clarify a question that had occupied physicists for a long time: does light consist of particles, as Newton believed, or of waves, as Christian Huygens claimed? Since Newton, as the father of mechanics, held greater authority, most eighteenth-century physicists followed his lead. It wasn't until Thomas Young conducted his double-slit experiment in 1802 that people were convinced that light had to consist of waves. And about a hundred years later, Einstein showed that light consists of particles, which raised the question of what light actually is.

The same is observed with electrons. Since these have a much smaller wavelength than light, one needs much smaller slits for this, but here, too, you can see a wave pattern on a photographic plate mounted behind the double slit. This experiment was first performed by Claus Jönsson in 1961 (Jönsson 1974).

The experiment becomes particularly peculiar when the intensity of the light (or electrons) is reduced to such an extent that only a single photon is put on the path. Again, an interference pattern appears on a screen placed behind the double slit. However, not in such a way that one first perceives a weak interference, which becomes stronger and stronger the longer one lets the experiment run. But one photon produces only one light spot on the screen. The next one produces another point of light at a different location. Thus, gradually more and more light spots appear, clustering in the maxima of a hypothetical wave, while hardly appearing in the minima. That actually particles are measured, was also to be expected, after all the wave function of the electron collapses (this appearance of the interference pattern was first observed using electrons, see Merli 1976), and the wave of the photons should behave similarly as the wave function of the electrons.

So, the question is: how does the photon know that it is going through a double slit and not a simple hole, and that it must produce an interference-like pattern on the screen and not just a bright spot?

One might assume that this means that a photon interferes with itself to create the interference pattern on the screen. But how should a quantum interfere with itself?

The representatives of the Copenhagen interpretation interpret this result in such a way that the quantum world can assume all states before the measurement, so that photons can go through both slits at the same time, albeit with different probabilities. After all, it is completely undetermined before the measurement. The question of its way is therefore not a meaningful question in the context of the quantum world.

In 1961, Claus Jönsson performed the first experiment showing interference of electrons at a double slit.

In 1976, Merli was able to show that the interference pattern is gradually built up by individual electrons.

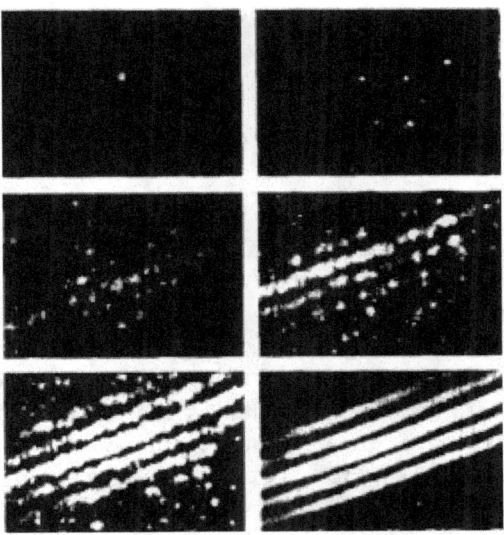

Another observation seems to support this interpretation: If one tries to find out, using a detector in front of the slit, which way the particle has taken through the slit, then immediately the interference pattern disappears and one sees only bright spots behind the double slit, which one would expect from particles. This observation can be explained by Bohr's idea of complementarity: You measure either particle or wave, but never both together. If you measure the path, then you measure the location of the particle and no interference occurs. If the path is indetermined, then the path of the particle "smears" and you measure the wave character of the quantum, resulting in the interference pattern on the screen.

The double-slit experiment seems to prove that a quantum can assume all states before the measurement and therefore can interfere with itself. But then a collapse of the wave function must occur during the measurement, which is instantaneous and thus violates the theory of relativity. And it still remains unclear, what a quantum, which is "smeared" over the whole world, actually looks like.

3.5. The Delayed Choice Experiment

The idea that, depending on the measurement, a quantum manifests itself sometimes as a particle (if you know the way) and sometimes as a wave (if you don't know the way), has a bizarre consequence, which can be observed in the so-called "delayed choice experiments".

In the late 1970s, the physicist John Wheeler wondered when in the course of the experiment it was determined whether the particle or the wave property would be measured (Wheeler 1978 and Wheeler 1983). He suggested an experiment with an interferometer, in which a light source first shines onto a beam splitter. This lets part of the light through to the right and deflects part of the light upwards. A mirror above the beam splitter reflects the light to the right, a mirror behind the beam splitter reflects the light upwards so that the light rays cross at the corner of the rectangle opposite the beam splitter.

Wheeler's Delayed Choice Experiment: A photon goes through the first beam splitter. One can now measure the path (above) or the interference. The type of measurement is chosen by introducing a second beam splitter after the photon has passed the first beam splitter.

With this setup, you can measure the path of the particle. You can attach a detector above or behind the light beam, which measures the particles that have either passed through or been reflected by the beam splitter, and thus have each taken a different path.

Now a second beam splitter is placed in the area diagonally opposite the first beam splitter, where the two light beams cross at the end of the setup. This second beam splitter can transmit the light beams coming from the side or from below or reflect them onto the other path. With this setup, you lose the path information – and you measure interference and the wave character of the quantum (but again the wave is only shown by the fact that the particles spread out in a wave-like pattern).

But when is it determined whether one is measuring particles or waves? The measurement starts when the quantum enters the interferometer. If a second beam splitter is not installed, then the particle character is measured. The quantum must therefore "decide" to take one of the two paths when entering the interferometer. If a second beam splitter is installed, interference occurs, so the wave must have taken both paths instead of just one – otherwise there could be no interference. The decision whether to measure the wave or the particle must therefore be made when the quantum enters the interferometer.

But what if you start with the setup to measure a particle, the quantum entered the interferometer, and then you change your mind and bring in the second beam splitter? The quantum was to be measured as a particle and took one of the two paths. The second beam splitter should therefore have no influence on the measurement, because the decision what to measure has already been made. But it actually has an influence: an interference pattern is measured as if the quantum had taken both paths (Jacques 2007).

The same applies vice versa: If one starts with the setup to measure the wave character and then removes the beam splitter after the quantum has entered the interferometer, then one measures a particle as if the wave had changed its mind to take only one path instead of both. The whole thing gives the impression that the decision made later (the "delayed choice") what to measure influences the path of the quantum – in a retroactive way.

The Copenhagen interpretation simply accepts this, since it assumes anyway that quanta are indetermined before the measurement. The actual measurement is ultimately done in the detector. That a quantum can be both a particle or a wave is possible until you actually carry out the measurement. What ingenious manipulations you do before the measurement therefore does not matter at all. But if the quantum world is completely undetermined before the measurement – how can you be able to set up physical concepts which allow you to unite the quantum world with the theory of relativity? If the Copenhagen interpretation were correct, then there would be no possibility to unite these two theories, because you cannot make any statement about the quantum world.

If you assume, as Wheeler did, that statements about the property of a quantum can also be made before the measurement in the detector, then you get a paradox, because the measurement influences retroactively the path of the quantum before the measurement. It does not seem to be possible to give a logical explanation about the quantum world.

So, the superpositions and the wave-particle duality make the quantum world a rather bizarre world. But it gets even more bizarre.

3.6. The EPR-paradox

The paradoxes that we have discusses so far are based on the fact that quanta sometimes show wave and sometimes show particle properties, and on the fact that the Schrödinger equation is linear, i.e. that a sum of solutions of the Schrödinger equation is again a solution of the Schrödinger equation, thus superpositions occur. This was interpreted in such a way that a quantum can assume all possible states before the measurement – at the same time. During the measurement all but one possible states disappear and only one remains, which according to Einstein is in contradiction to the theory of relativity.

It was also Einstein who pointed out another paradox of quantum mechanics; in his search for a better understanding, he had evidently delved deeper into the bizarre world of quanta than many of his colleagues. This paradox is based on the phenomenon that the Schrödinger equation is not defined in position space but in configuration space.

Newton's equations describe the motion of a particle in position space. One knows the coordinates of the particle and can then describe its movement in the coordinate system of the position space.

The Schrödinger equation, on the other hand, supplies a wave function for all particles involved, i.e. for the configuration of these particles in relation to one another. One no longer describes the movement of a particle in position space, but the behavior of an ensemble of particles in configuration space.

However, this has a remarkable consequence, as Einstein recognized: If the state of a particle changes, this has an instantaneous effect on the state of the other particles, because the wave function acts directly on all particles.

Usually, this interaction is so small that it can be ignored and the particles can be treated independently from each other. But this is not always the case. Sometimes, the particles are deeply entangled and have a direct influence on each other.

Schrödinger called this phenomenon, that particles are not independent of each other but influence each other instantaneously, entanglement (Schrödinger 1935a), shortly after Einstein had pointed out this "spooky action at a distance" in his famous paper in 1935.

Einstein had published the paper together with his colleagues Boris Podolsky and Nathan Rosen, which is why the paradox is also known as the EPR paradox after the first letters of the last names of the authors involved (Einstein 1935).

In the paper, the authors argued that quantum mechanics according to the Copenhagen interpretation is incomplete. The starting point is two entangled particles that are far apart, the first on Earth, for example, and the second on Jupiter. Now one carries out a measurement of the impulse of the first particle, and measures the position of the second particle. According to Heisenberg one cannot know both information exactly as there is a minimum uncertainty. If one parameter is known exactly, the other one is completely unknown.

But the particles are entangled with each other. The Schrödinger equation supplies a sum value for the momentum. If you measure the momentum of the first particle, you can calculate the momentum of the second particle exactly. If you measure the position of the second particle, then you can calculate the position of the first particle exactly. The position and the momentum of both particles are therefore known exactly – in contradiction to the basic assumption of the Copenhagen interpretation that both parameters are complementary and cannot be measured at the same time.

The EPR paper presented two ways out of this dilemma: One could assume that the two parameters are undetermined before the measurement and are only determined during the measurement, i.e. they are influenced by the measurement. If the momentum of the first particle is measured, then the momentum of the second particle is determined instantaneously. However, since the distance between the two particles can be arbitrarily large, the information about the momentum of the first particle would have to reach the second particle faster than the speed of light so that the second particle adjusts itself accordingly. However, this is forbidden by the theory of relativity. Einstein therefore considered this way out to be impossible. If there were such an action at a distance, it would be, in Einstein's words, "a spooky action at a distance".

Then only one solution remained, which Einstein called "physical reality": The two parameters are not determined only by the measurement and are completely undetermined before, as the Copenhagen interpretation assumes, but the two parameters are determined from the beginning and only unknown until the measurement is performed. Only when you carry out the measurement, you find out these parameters. Before that, they are "hidden variables". These hidden variables determine the behavior of the quantum particles. Since the Copenhagen interpretation does not know these variables, Einstein's conclusion was that this interpretation of quantum mechanics must be incomplete.

The EPR paradox: Can measuring the parameter of a particle on Earth affect the corresponding parameter of an entangled particle on Jupiter instantaneously? Or are there hidden parameters that determine the behavior of the particles in advance?

Einstein's thought experiment raises fundamental questions about the philosophy of quantum mechanics. However, this question was too esoteric for most physicists, and there was no way to realize this thought experiment. The EPR paper therefore went largely unnoticed in the first few years after its publication.

This did not change when David Bohm published a version of the EPR paradox that was easier to measure (Bohm 1957). He used the spin of electrons for this. If two entangled electrons have a total spin of zero, and one spin is measured as -½, then the other electron must have a spin of ½. The same applies to the polarization of light. Spin and polarization were easy to measure, but in the 1950s there was not much interest in philosophical questions. The EPR paradox was just another paradox of quantum mechanics that physicists believed to show how peculiar the quantum world was. One could calculate the behavior of the particles, but not understand them. And that's what the physicists did, and that's how modern computer chips, lasers and superconductors came about. Quantum theory was obviously very successful, even if all you had were mathematical formulas that defied any clear understanding.

4. Interpretation of quantum mechanics

Physics makes a model of the world, it tries to explain the world, which is why it sometimes has to simplify it. Newton described gravity as a force that attracts bodies to each other, even though he could not explain how this force should work. Einstein later explained gravity as distortions of space-time produced by the masses in space, although it is not entirely clear what is actually meant by "space".

These models are extremely successful, but they must not be confused with reality. An artist paints a picture of a tree, but the picture is not a tree. The Belgian painter René Magritte painted a picture of a pipe and labeled it "Ceci n'est pas une pipe" (This is not a pipe) to make this difference clear. The models help us to understand reality, but they are only interpretations of the world. One must not confuse the image with reality. Yet that seems actually to be the case in quantum mechanics. Physicists have so internalized the Copenhagen interpretation that they believe it describes the quantum world without any ifs and buts. We will see in the following, that this is not the case. For this purpose, we will have a closer look at how the Copenhagen interpretation actually came about.

4.1. Lifting the curtain

Einstein had shown in 1905 that light exists in the form of energy quanta – but at the same time, according to Maxwell's laws, it is also a wave. As part of his doctoral thesis, Louis de Broglie examined the wave and particle properties of matter more closely and made two amazing observations, which he described in detail in his doctoral thesis from 1924 (de Broglie 1925).

One observation concerned a contradiction in Einstein's theory of relativity. We will go into more detail about this theory later, but the basic assumptions of the special theory of relativity are simply explained: For all observers, the speed of light in a vacuum is the greatest speed, no matter how fast an observer is moving, and physical measurements must provide the same result for all observers, otherwise an observer could see how fast he is moving and the reference systems would not be equal. This has the strange consequence that time passes more slowly in a fast-moving system, and at the same time the mass of a body increases. When a system approaches the speed of light, time passes infinitely slowly and the mass of a body becomes infinitely large. Both of course is nonsensical, which is why no body can move faster than the speed of light (and only massless bodies like photons can move with the speed of light). In addition, according to Einstein, mass is proportional to energy, and the nice formula is valid: $E=mc^2$, where c is the speed of light.

Planck had determined that the energy E of a photon is proportional to its frequency v, with the constant of proportionality being Planck's quantum of action h. I.e. it holds: $E=hv$.

De Broglie had noticed an inconsistency in this picture:

If a body moves faster, its mass increases. Since mass and energy are equivalent, this means that also its energy increases and thus its frequency. So, on the one hand, you get that the frequency of a body increases with its speed.

If a body moves, then, however, also time stretches. The frequency is the inverse of the time. If the time becomes larger, then the frequency becomes smaller. So, you also get the exact opposite result.

If you start from energy, frequency must increase with speed, if you come from time it must decrease, as de Broglie had observed. How can both be right at the same time?

De Broglie proposed a special interpretation for this. According to it, the formula that frequency decreases with speed (which had been derived directly from the theory of relativity) describes a particle that is moving at a certain speed and vibrates at that frequency. The speed of the particle is less than the speed of light. The result that the frequency increases with speed comes from the energy-frequency relation, i.e. from the investigation of waves that, according to Planck and Einstein, show particle character. This provides waves with a phase speed that is greater than the speed of light. This is not a problem at first, since no information can be exchanged with the phase alone, for this you need the amplitude of the wave (because the energy is linked to the amplitude) or the group velocity.

De Broglies investigations brought him on the fact that there is a particle which can move at most with the speed of light and a phase wave which reaches speeds which are clearly above the speed of light.

De Broglie assumed that a particle is guided by a wave field. The path of the particle is thus determined, but an interference pattern still occurs as the guidance of the particle is realized by a wave. The "duality" of the quantum world actually reflects two interconnected properties of the quantum world.

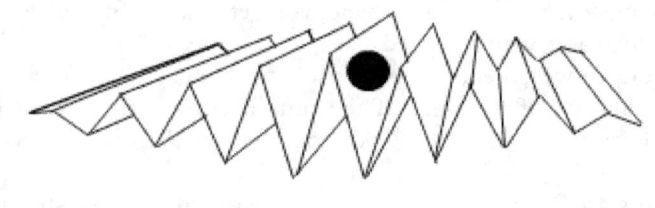

The phase wave and the particle oscillate with the same frequency. The particle moves with finite speed, the phase wave is faster than the speed of light. De Broglie now assumed that the phase wave propagates practically instantaneously in space and guides the particle. The phase wave is a guiding wave, which guides the particle on paths which do not resemble the straight-line paths of Newtonian mechanics, but are "wave-like" – and this although no further force acts (for the exact mathematical derivation see appendix A).

According to the extremal principle, an object takes the path with the shortest time. In order to determine this, one assumes that the light, e.g. when refracted on a glass surface, could take countless paths. Then, according to Fermat, the phases are summed up and the minimum path is calculated; for near the minimum the phase of the light changes only slightly, this path is therefore amplified; paths that are further away are weakened due to the large phase difference, so that the optimal path can be calculated.
This summation over phase is also the basis for Feynman path integrals, which are used in quantum electrodynamics.

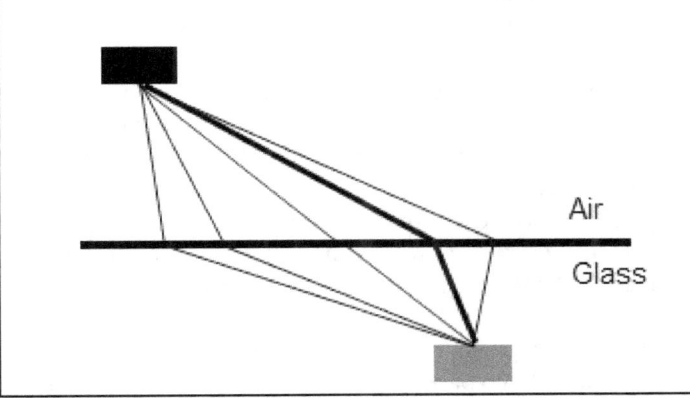

Air

Glass

De Broglie supported this claim with his second observation, namely that Fermat's principle for the propagation of waves and

Maupertuis' principle for the propagation of particles are equivalent.

Both principles are based on the extremal principle, the principle of least action. According to Fermat, the path of light is obtained by adding up all possible phases. The phases are slightly different for each way the light can go, as the length of the way differs. However, at the minimum, where the time to take the path is shortest, the phases change only slightly. This is the path the light will take, as this is amplified, whereas area with a large change in phase away from the minimum will cancel out.

Starting with Fermat's formular and using the phase velocity de Broglie had obtained from the equivalence of energy and frequency, as well as the equivalence of energy and matter, de Broglie obtained the formula for the principle of least action that Maupertuis had set up for particles.

The principle of the least action describes how a particle moves through space. If the phase velocity of the wave corresponds to the faster-than-light phase velocity which de Broglie had obtained from the equivalence of energy and frequency, then the principles of least action for particle and wave are identical. The "something" that leads the particle through the space, therefore, had to be the phase wave or guiding wave according to de Broglie. And these considerations should be valid for light as well as for normal particles, as de Broglie claimed.

When he presented this audacious theory in his doctoral dissertation, his thesis advisor, Paul Langevin, was doubtful as to whether this theory, that matter particles could also exhibit wave properties as they were guided by a guiding wave, could actually make sense. He therefore sent it to Albert Einstein, asking for his opinion. Einstein was impressed and said that de Broglie had "lifted a corner of the great curtain". De Broglie then received his doctorate.

De Broglie's idea spread quickly. Einstein had shown that light waves could also have particle properties, de Broglie had shown that matter particles could also have wave properties. However, de Broglie's idea was often abbreviated. De Broglie assumed that particles are guided by a wave, but he was understood to have said that quanta could also have wave properties in addition to the known particle properties. Davisson and Germer then demonstrated these wave properties in 1927 by shooting electrons at a nickel crystal and observing interference.

This could now also explain Bohr's atomic model. Bohr had assumed that electrons can only move around the atomic nucleus in certain orbits. If electrons have wave properties, then the waves of electrons moving around the nucleus interfere. This interference is mostly destructive, the waves cancel each other out. At certain distances from the atomic nucleus, however, the length of the waves is just so long that wave crests meet wave crests, and the waves interfere constructively. Electrons can stay here. Only these orbits can exist in an atom.

Although de Broglie had assumed that particles are guided by waves, he could not give a wave equation. However, Erwin Schrödinger succeeded in doing this – with which he opened the floodgates to the Copenhagen interpretation, although he himself rejected this interpretation.

4.2. Wave equation and Copenhagen interpretation

Erwin Schrödinger published his wave equation in 1926 (Schrödinger 1926a, 1926b). Without further assumptions, this wave equation supplies the correct energies for the electrons in the hydrogen atom, for which Bohr had to make special assumptions that he could not explain. The correct energies result solely from the fact that electrons move like waves and the fact that the electrons are "trapped" in the electric field of the atom.

However, the question arose as to what this wave equation actually describes physically. Electrons aren't just waves, they are also particles, so they are concentrated in one place, at least when you measure them. Schrödinger tried to save the particle picture by assuming that this arose because the wave was a wave packet. Wave packets are locally concentrated, while a free wave propagates everywhere in space. However, a free wave consists of a single wave with one wavelength, whereas a wave packet is a superposition of multiple waves that interfere in such a way that there is a maximum that quickly decays to either side. First of all, this is not a problem for the Schrödinger equation, which allows a superimposition of waves. But it is a problem from a physical point of view: Waves with different wavelengths travel through a medium at different speeds, they experience dispersion. As a result, the wave packet becomes wider and wider over time until it eventually completely disintegrates. If particles were wave packets, they would only have a certain lifetime. Should one now assume that this "dispersing" of the wave packets does not exist for particle waves? But there would be no other reason for this assumption except to save the hypothesis of wave packets. So, the wave of the Schrödinger equation had to have a different meaning. Complicating matters further is the fact that the waves in the Schrödinger equation are generally not physical waves. The Schrödinger equation is a complex equation. It contains the imaginary number i, i.e., the root of -1. Thus, the waves as solutions of this equation are also usually complex wave functions. However, complex wave functions do not describe real waves. Thus, the Schrödinger equation yields waves that cannot exist in the physical world at all. So, what are these waves?

Max Born provided the already mentioned solution for this: The wave function does not describe a real, physical state, only the absolute square of the wave function provides a description for the probability with which a particle assumes a certain state. If you form the absolute square of a complex number, then you receive a real number, and this is physically meaningful. Instead of concrete particles you only know the probability with which a particle can be found somewhere.

The Schrödinger equation thus provides distributions of the probability that particles are in a certain area around an atom, the so-called orbitals. Depending on the orbital angular momentum of the electrons, the orbitals are sometimes circular, sometimes dumbbell-shaped or have other shapes. This provides information about where to find an electron with high probability.

The Schrödinger equation is:

$$i\hbar\frac{\partial}{\partial t}\Psi\,(r,t) = H\Psi(r,t)$$

with
$\hbar = h/2\pi$,
H the Hamilton operator describing the energy of the system, and
$\Psi(r, t)$ the wave function as a function of location r and time t.

Since the Schrödinger equation is a complex equation, the wave functions are usually complex functions that do not describe any physical reality.

Physical relevance has the square of the absolute value of the wave function:

$$|\Psi(r,t)|^2$$

It describes the probability of finding a particle in a specific place at a specific time.

The Schrödinger equation is also a linear equation. This means that a sum of two solutions is also a solution of the Schrödinger equation. This leads to the superposition of states and the strange behavior that a cat can be mathematically dead and alive at the same time.

Also, the Schrödinger equation does not describe the behavior of one particle alone, but it describes the behavior of all involved particles, so it is not defined in the location space of one particle, but in the configuration space which contains the coordinates of all particles. Therefore, the Schrödinger equation allows the strange behavior that the state of one particle can instantaneously affect the state of all other particles, the entanglement that Einstein pointed out in the EPR paper.

Superposition and entanglement are first of all consequences of the Schrödinger equation, which is a model of quantum mechanics. But because the Schrödinger equation describes the states of the particles so well, one considers these consequences to be real. There must be superpositions and entanglements of the quanta in the world.

At the same time, quanta sometimes show wave and particle properties, depending on which measurement setup is used. How do you reconcile all this information? How should one imagine the quanta that move through the world both as waves and as particles at the same time here and there and then also entangled with each other?

The Copenhagen interpretation, which was mainly advocated by Niels Bohr, but which was followed by almost all physicists of the time apart from Einstein, Schrödinger and de Broglie, was pragmatic: Not at all.

Ultimately, we can only know what we can measure. We can speculate about what is before the measurement, but we cannot know it. A quantum is completely undetermined before the measurement.

The assumption sounds absurd: Everyone assumes that the moon orbits the earth even when nobody is looking. That may be so in the macroscopic world, say the supporters of the Copenhagen interpretation, but it is not so in the quantum world. A quantum is not defined before the measurement. It is only a probability. The world is no longer deterministic, but non-deterministic. Since it is not determined before the measurement, it no longer makes sense to speak of cause and effect.

According to the Copenhagen interpretation, a quantum can be a wave or a particle, and it can be anywhere with a certain probability. Its character is only defined by the measurement. Before that it is indetermined. The quantum world is therefore non-deterministic.

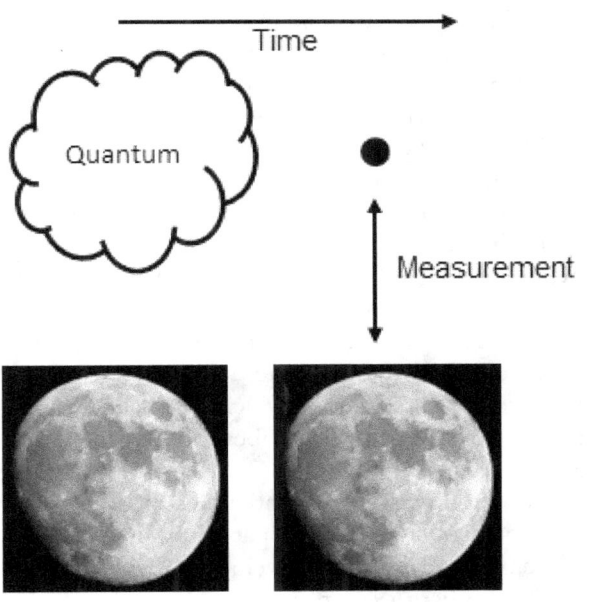

Time

Quantum

Measurement

This is different for classic objects like the moon. These are also determined before a measurement.

This probability is therefore stronger than the probability in thermodynamics. In thermodynamics you only calculate statistically because you do not know the individual parameters of billions of particles. Therefore, you use statistics and probability.

Theoretically it would be possible to collect this data. In practice, however, it is not feasible.

In quantum mechanics, the whole thing is one step more complicated according to the Copenhagen interpretation. Here it is in principle impossible to describe exactly the state of a particle before the measurement. Probability is part of the quantum mechanical world, as the Copenhagen interpretation claims, and is not only used because you do not have all the information. An electron is therefore smeared over the whole space, its states overlap in the whole space, and the Schrödinger equation can only tell us the probability to find a particle at a certain place. Only at the measurement the electron will manifest itself at a certain place. This is also required by Heisenberg's uncertainty relation. Originally, Heisenberg had explained it "classically" with the example of how you determine the location of an electron. For this you need a photon. If you want to determine the place exactly, then it must be a high-energy photon with small wavelength. However, if it collides with an electron, the photon is deflected, but so is the electron. When the location is measured, the momentum of the electron is thus changed. You cannot determine both parameters at the same time.

After this clear explanation, which Einstein and Schrödinger would have approved of, Heisenberg went one step further. He wrote:

> If one admits that discontinuities are somehow typical for processes in very small spaces and times, then a failure of the terms "place" and "velocity" is even directly plausible (Heisenberg 1927, p. 173).

In the quantum world, place and velocity or momentum cannot be measured at the same time, not only because the measurement of one parameter changes the value of the other parameter, but the parameters place and velocity are no longer meaningful terms in the quantum world. Only when you measure something you can make a statement about the quantum world. Before that, the system is undefined.

Bohr introduced the concept of "complementarity". A quantum is not a wave or a particle, but can be described as a wave in one context and as a particle in another. A wave provides the momentum of the quantum but leaves its location undetermined, a particle provides the location but says nothing about the momentum. This is how the concept of complementarity fits together with the uncertainty principle.

The Copenhagen interpretation seems to describe the measurements sufficiently (it deals only with them and leaves open what happens before), however, it leaves a bad taste because a quantum is undefined over large parts of its life, as if the quantum exists only as a possibility and only the measurement calls it into being. In the classical world, one could describe the behavior of particles even if one did not measure them, and arrive at correct values for measurement. In the quantum world, the behavior of the quantum is said to be indetermined and can only be described using probabilities until the quantum is then measured and reveals its properties. Einstein and others could not come to terms with this idea during all their lives. Einstein summed up his reservations about this notion of the Copenhagen interpretation by saying "God doesn't play dice!" He believed that it was due to missing information, which would later be called hidden variables, that only probability statements could be made. He did not believe, however, that in the quantum world you can only make probability statements, because quanta are not determined before the measurement, that the probability statements in the quantum world are therefore of fundamental nature.

However, most physicists accepted the Copenhagen interpretation or did not care about what the quantum world really looks like, after all the formulas, as incomprehensible as they may be from a physical point of view, gave quite precise results that agree well with the observation. The question of what the quantum world really looks like was hardly asked after the 1930s. They acted according to the credo "Shut up and calculate!" and were happy with the nice results.

However, it was regretted that this interpretation of the quantum world, which left the quanta largely undefined, could not establish a connection to the theory of relativity; because if objects are undefined, then one cannot make a physical statement and develop a physical concept that would allow connecting the quantum world with the world of the theory of relativity. The "theory of everything" must remain a dream if the quantum world is incomprehensible.

But that's not entirely true either. A few physicists continued to try to understand the quantum world and to find an interpretation that was "more intuitive" than the Copenhagen interpretation.

4.3. The de Broglie-/Bohm-interpretation of quantum mechanics

The first, of course, was Louis de Broglie. Initially, he had not succeeded in giving a formula for his guiding wave, but with the Schrödinger equation, there was now an equation for a wave, even though the waves obtained by it had no direct physical meaning, but first had to be squared to obtain a probability statement. But de Broglie was now able to show that the wave obtained from the Schrödinger equation determined the velocity of the particles in his model of the guiding wave, and that the velocity of a particle is proportional to the spatial derivative of the phase (de Broglie 1927).

But by this time, physicists had become accustomed to calculating with wave functions alone. According to de Broglie's approach, one first had to calculate the wave function to obtain the guiding field, and then calculate the path of the particle. This sounded like additional work that somehow did not seem to add any value. Therefore, de Broglie's approach was forgotten, and de Broglie himself did not pursue it any further in the next years.

This changed in the 1950s, when David Bohm rediscovered de Broglie's idea. The starting point was a textbook on quantum mechanics that Bohm wrote in the 1940s. In it he presented the Copenhagen interpretation in detail. But the longer he occupied himself with this interpretation, the more uncomfortable he felt with it. He had the impression that it couldn't be correct. He therefore set out to find a new interpretation that seemed to be more consistent in his view. In doing so, he rediscovered de Broglie's interpretation of the guiding wave (Bohm 1952a, 1952b). However, he did not focus on the fact that a wave guides the particle, but that there is a quantum potential (derived from the wave of the Schrödinger equation as well) that influences the motion of the particle (for the derivation see appendix B).

According to the de Broglie-/Bohm-interpretation of the quantum world, a particle is guided by a wave or a quantum potential. In this way, the measurement points are distributed as if the quantum were a wave. Which path the particle takes, here the possible paths with a double slit, depends on the initial position of the particle.

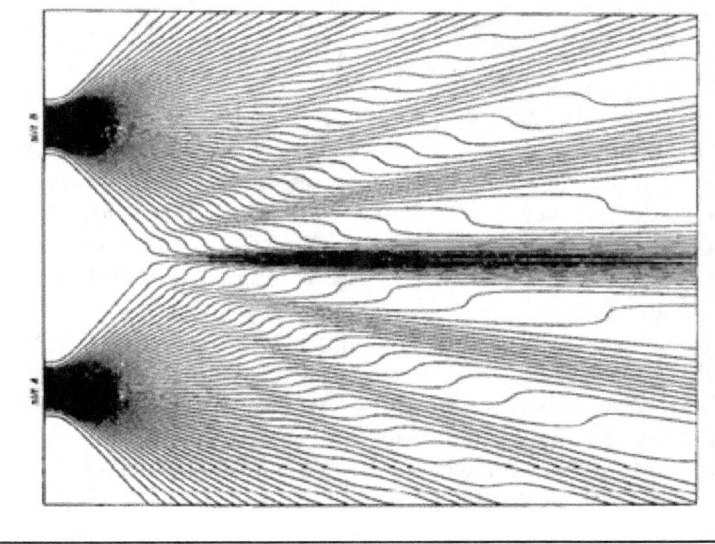

Bohm's work made de Broglie revisit his original theory, but the de Broglie-/Bohm-interpretation was still ignored by other physicists, even though it provided the same predictions as standard quantum mechanics, because they had two problems with it: First, compared to Schrödinger's wave mechanics, the theory became more complicated because particle trajectories had to be calculated in addition to the wave. The fact that Schrödinger had originally removed the particle paths from the de Broglie theory had already been forgotten at that time. And on the other hand, it became clear from the formulas of the de Broglie-/Bohm-theory that the wave changes instantaneously. As the phase of the wave influences the path of the particle, this theory would allow changes beyond the speed of light.

So-called non-localities, where something moves faster than the speed of light, also exist in the Schrödinger equation, which after all was the basis for the de Broglie/Bohm interpretation, just think of the collapse of the wave function, but in the case of the Schrödinger equation alone one took refuge in the Copenhagen interpretation: Before the measurement, a particle is completely undefined, so you cannot say whether something has changed non-locally at the time of the measurement. However, this excuse was not possible for de Broglie/Bohm. Here was no other possibility than the conclusion that something had to move faster than the speed of light. And this was probably also the reason why Einstein, who himself also worked on the idea that particles can be guided through a field, found de Broglie's theory good in the beginning, but did not support it completely; for this theory exhibited non-localities that seemed incompatible with the theory of relativity.

The obvious non-locality of Bohmian mechanics, as it is often called, because de Broglie's part was forgotten, was therefore a reason why many physicists did not take it seriously. However, the de Broglie-/Bohm-interpretation of quantum mechanics has a great advantage over the Copenhagen interpretation: With it, many paradoxes vanish into thin air, and one realizes that these paradoxes are not paradoxes of the quantum world, but paradoxes of a special interpretation of the quantum world.

5. Paradoxes revisited

So far, we have looked at the known paradoxes of the quantum world through the eyes of the Copenhagen interpretation, which is considered by most physicists to be the standard interpretation of the quantum world. Now let us look at these paradoxes from the perspective of the de Broglie-/Bohm-interpretation. Since the de Broglie-/Bohm-interpretation delivers the same results as the standard interpretation of quantum mechanics (Dürr 1992, 1993), it is also a possible interpretation. It will be shown that many paradoxes disappear completely. They are therefore not paradoxes of the quantum world, but paradoxes of a special interpretation of the quantum world, namely the Copenhagen interpretation. Only one paradox persists from the point of view of the de Broglie-/Bohm-interpretation – which could mean that this is actually a "problem" of quantum mechanics and not of a special interpretation.

5.1. The collapse of the wave function

According to the Copenhagen interpretation, a quantum exists before measurement as a superposition of possibilities. It is just as likely to be at the location of the future measurement as it is on Jupiter, albeit with a significantly lower probability. During the measurement, the wave function collapses, so it no longer exists instantaneously on Jupiter, although it would take some time for light to travel from Earth to Jupiter. The collapse of the wave function would thus contradict the theory of relativity. The Copenhagen interpretation circumvents the problem by claiming that a quantum is undefined before measurement, which is why one cannot deduce that any rule is violated in the measurement.

After all, the particle is only determined at the measurement, before that it is only a possibility. However, this possibility should be distributed over the entire space, so that these arguments are not a real solution to the problem.

This problem does not exist in the de Broglie-/Bohm-interpretation. The particle is guided by the wave function, the square of the absolute value of which indicates where the particle is and with what probability. If you carry out a measurement, you measure the already existing particle. There is no collapse of the wave function in the de Broglie/Bohm interpretation, since the wave function only describes the guiding wave for the particle, but is not itself the quantum. Thus, there is no corresponding paradox. The collapse of the wave function is not a paradox of the quantum world but of the Copenhagen interpretation.

5.2. Schrödinger's cat

With his example of the cat Schrödinger wanted to show how nonsensical the assumption is that in the quantum world a quantum can assume several states at the same time before the measurement. In the world we know, this leads to the absurd consequence that a cat can be dead and alive at the same time and only assumes one of the two states when we open the box and look.

In fact, however, we assume that a cat is either dead or alive, but we just don't know this until we open the box and look.

The supporters of the Copenhagen interpretation say that the quantum world behaves just differently. Here, a quantum can assume all possible states until the measurement is made and the wave function collapses.

In the de Broglie-/Bohm-interpretation, this paradox does not exist. Here a particle is well defined before the measurement and it exists in exactly one state – we are just not able to say which one before the measurement is done, just as we cannot know exactly the behavior of the particles of a gas and therefore have to help ourselves with statistical statements. The paradox of Schrödinger's cat is therefore also a paradox of the Copenhagen interpretation and not a paradox of the quantum world.

5.3. Wigner's friend

The paradox of Wigner's friend dissolves just as easily in the context of the de Broglie-/Bohm-interpretation. According to the Copenhagen interpretation, the wave function had collapsed for Wigner's friend after the measurement of the quantum, but for Wigner, who did not yet know the result of the measurement, it still existed. Both at the same time is simply not possible, and shows how bizarre the quantum world is. In fact, it just shows how bizarre the Copenhagen interpretation is.

Because according to the de Broglie-/Bohm-interpretation there is no paradox. The particle is well defined before the measurement, but its parameters are not known because they have not yet been measured. Wigner's friend now carries out the measurement. Since he has not yet communicated them to Wigner, Wigner cannot know what these parameters look like. But they are also well defined at this point. They just haven't been told to Wigner yet.

The situation is comparable to measuring the depth of a well. An employee measures the depth and thus knows the value. The site manager does not know the value at this point. Nevertheless, the depth of the well is also well-defined for him, and as soon as the employee has told him the depth, he also knows the exact value.

Therefore, also the paradox of Wigner's friend is not a paradox of the quantum world, but only a paradox of the Copenhagen interpretation.

In the Copenhagen interpretation, a quantum is a wave and a particle and can be anywhere at the same time. This leads to numerous paradoxes, such as the collapse of the wave function and Schrödinger's cat, which can be alive and dead at the same time.

In the de Broglie-/Bohm-interpretation, a particle is guided by a wave (or a potential). Many paradoxes of quantum mechanics thus vanish into thin air - and one recognizes that these are not paradoxes of the quantum world, but paradoxes of the Copenhagen interpretation.

5.4. The double-slit

Let's now turn to the interference experiment of the double slit, which Feynman called *the* riddle of quantum mechanics. As long as countless photons pass through the double slit at the same time, one has no problem imagining some kind of interaction between the photons that leads to an interference pattern which is observed on a screen behind the double slit. But as soon as only a single photon flies through, this becomes problematic: How should a photon interact with itself so that it lands on the screen as if there were an interference phenomenon?

The Copenhagen interpretation explains this by saying that a photon can assume all possible states and can therefore also interfere with itself. The double-slit experiment is therefore the convincing proof that the Copenhagen interpretation is correct. But how should one imagine that an actually indivisible photon is divided into billions of subparticles that interact with each other? No wonder Feynman called this *the* riddle of quantum mechanics. However, the puzzle dissolves in the de Broglie-/Bohm-interpretation. The guiding wave (de Broglie) or the quantum potential (Bohm) defines the possible way which a particle can take. The guiding wave is influenced by the double slit as if it would interfere with itself. Thus, there is a guiding wave which sends the particle on its path which leads to an interference pattern – and the particle ends up with high probability in the maximum of a theoretical interference pattern. In this interpretation, one does not have to assume that the particle interferes with itself, since it is guided on its path by a guiding wave.

If you want to investigate the path of the particle, e.g. by covering a slit, then the guiding wave changes accordingly; because now there is no second slit with which interference can occur, and the particle is sent on a path which resembles the straight-line path of a particle.

In the Copenhagen interpretation the double slit is *the* riddle of quantum mechanics, in the de Broglie-/Bohm-interpretation the behavior follows easily from the behavior of the guiding wave, without having to assume that the photon would interfere with itself. The photon moves through space undivided and lands on the screen as a particle – where it is then also measured as a particle, without any collapse of the wave function. The problem of the double slit is also a paradox only in the Copenhagen interpretation.

5.5. The Delayed Choice Experiment

In the delayed choice experiment, the decision whether to measure the wave character through interference or the particle character through path information is made after the quantum has entered the measuring apparatus, the interferometer. One has the impression that this decision can retrospectively determine how the quantum behaves.

The Copenhagen interpretation circumvents the problem by saying that a quantum is completely undetermined anyway before the actual measurement, which takes place in the detector and not when entering the interferometer. There is no point in worrying about the appearance of an object that is indetermined.

The de Broglie-/Bohm-interpretation explains the different behavior by the fact that changing the measurement setup also changes the guiding wave (the quantum potential). If you want to measure an interference and setup the measuring apparatus accordingly, then the guidance wave is formed in such a way that you measure an interference. If you want to measure the particle character, by measuring the way of the quantum, then you change the measurement setup accordingly, and another guiding wave is formed, which allows the measurement of the particle character.

If you change the measurement setup while the quantum moves through the interferometer, then the guiding wave changes accordingly. If you started with a guiding wave which would have produced an interference behavior then you have a guiding wave which measures the way and shows the particle character after you have changed then measurement setup.

But: This change in the guiding wave would have to happen quickly, very quickly. Maybe too fast? After all, the wave function, which ultimately describes the path of the particle, allows the guiding wave to change instantaneously. That was also a criticism of the de Broglie-/Bohm-interpretation. Does the de Broglie/Bohm interpretation not solve the riddle of the delayed choice experiment, but only replace it with another one? Or are we getting a glimpse of a truly fundamental feature of the quantum world?

5.6. The EPR-paradox

Einstein had described in the EPR paper the puzzle that the measurement on one particle would affect the properties of another particle at the other end of the world – instantaneously. Einstein thought this interpretation was nonsensical because it contradicted the theory of relativity, so there had to be another solution: The behavior of the particles had to be fixed before the measurement, there had to be "hidden parameters", quantum mechanics was incomplete.

The de Broglie-/Bohm-interpretation does not solve this paradox in the sense of Einstein, either. It assumes that the guiding wave, which controls the behavior of the particles, is non-local. So, it would also allow that something moves faster than the speed of light.

According to de Broglie/Bohm, the solution of the EPR paradox would be similar to the solution of the delayed choice experiment: Something moves faster than light. But this contradicts the theory of relativity.

Alternatively, there would of course be the possibility that the behavior of the particles is already defined before the measurement. In the EPR experiment, one measures the spin of a particle, but contrary to what the discussed quantum theories assume, the orientation is not influenced by the measurement (or, as in the case of the Copenhagen interpretation, defined by the measurement), but the orientation has been determined before the measurement.

Likewise, with the delayed choice experiment, one would have the impression that we determine in the course of the experiment which property of the quantum we want to measure, but in fact this is already determined beforehand, since the physical world, and this includes our behavior, is completely determined. We have the impression to decide, but the decision has already been made. Thus, there are hidden parameters that determine the behavior of a particle that we just don't know about yet.

However, one paradox remains regardless of the interpretation: the non-local influence between quanta.

Are there indeed non-local phenomena in quantum mechanics, as the formulas suggest, or is it rather that there are hidden variables that determine the behavior of quanta that we do not know yet, which is why we think that there is a non-local influence?

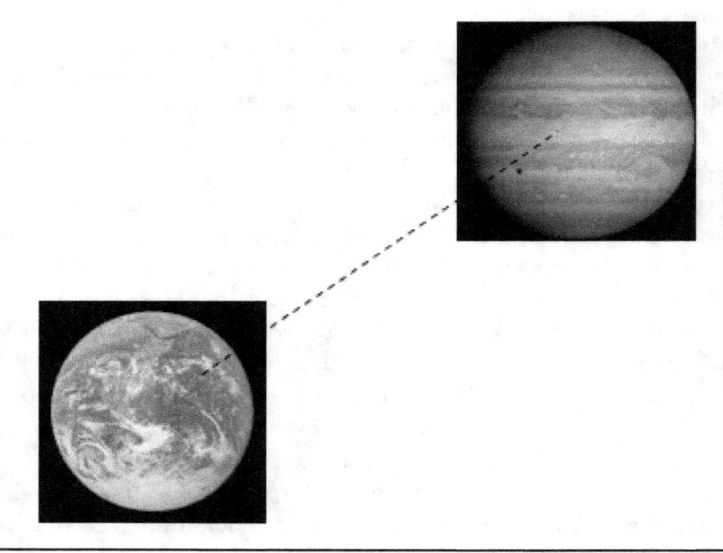

Our considerations have brought us to the point that we could unmask many paradoxes of quantum mechanics as paradoxes of the Copenhagen interpretation. But not all paradoxes of quantum mechanics are bogus problems generated by the Copenhagen interpretation. There seems to be something that is actually a paradox of the quantum world, that remains with us no matter what interpretation we use. We now have reached a crossroads, as Einstein recognized almost ninety years ago: Either it is indeed the case that the quantum world shows non-locality and thus potentially violates the theory of relativity, or we must assume that quantum mechanics is incomplete and there are "hidden parameters" that we just haven't found yet.

As fundamental as the question is for understanding the quantum world, it remained unanswered for decades.

6. Bell's inequality and non-locality

It wasn't until the 1960s that a physicist revisited the question Einstein had raised: Is quantum mechanics incomplete, and is the probability that we find in the quantum world just a consequence of the fact that we cannot measure the quantum world accurately, so that we have to work with statistics? In other words: Are there any hidden parameters? Or does probability in the quantum world go beyond what we know from our local environment, and are there non-localities in the quantum world?

The physicist John Bell wondered if one could somehow distinguish these two possibilities experimentally. He published his considerations in 1964 in the form of an inequality, the famous Bell's inequality (Bell 1964). Bell assumed that there were hidden parameters. Then he thought about what the probabilities should look like if you measure two out of three parameters.

Let's take the simple example of people. People can be tall or short (abbreviated a_t and a_s), blond (b_b) or dark (b_d), and man (c_m) or woman (c_w). Now we measure two of each of these characteristics. The number of tall blondes is equal to the number of tall blond men plus the number of tall blond women. If you omit one of the three features, then the number found is either equal to or greater than. Thus, the number of tall blondes is less than or equal to the number of blond men (which now includes all the short ones) plus the number of tall women (which now includes the dark-haired ones). So, it holds (if N is the number):

$$N(a_t, b_b) \leq N(b_b, c_m) + N(a_t, c_w)$$

This inequality holds if there are hidden variables. The parameters of the people with regard to size, hair color and gender are fixed before the measurement.

If you have a crowd of tall people and short people, of blondes and brunettes, of men and women, then the set of tall blondes is less than or equal to the number of blonde men plus the number of tall women. A similar relation in the quantum world describes Bell's inequality, which applies to hidden parameters, i.e. to characteristics that are defined before the measurement.

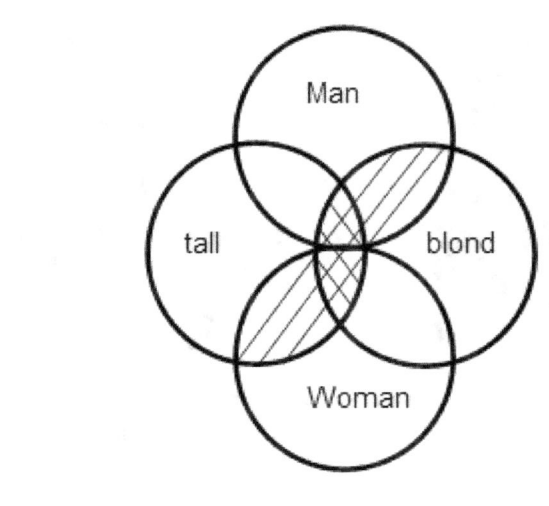

Bell considered the probabilities of measuring the spin of entangled electrons and obtained an inequality that describes a linear dependence of the measurement results on the angle of the measurement (for the exact derivation, see appendix C). However, it follows from quantum mechanics that the measurement results must be proportional to the cosine of the measurement angle. Thus, the probabilities should be different with the exception of the angles 0°, 90° and 180°, where classical probability and quantum mechanics give the same results. If the laws of quantum mechanics differ from those of the classical world, then one should be able to measure a violation of Bell's inequality.

Let's look at the spin measurement on an electron. If spin +1/2 is measured on the first particle, then the entangled particle has spin -1/2. The probability of measuring one of the two spins is 50%.

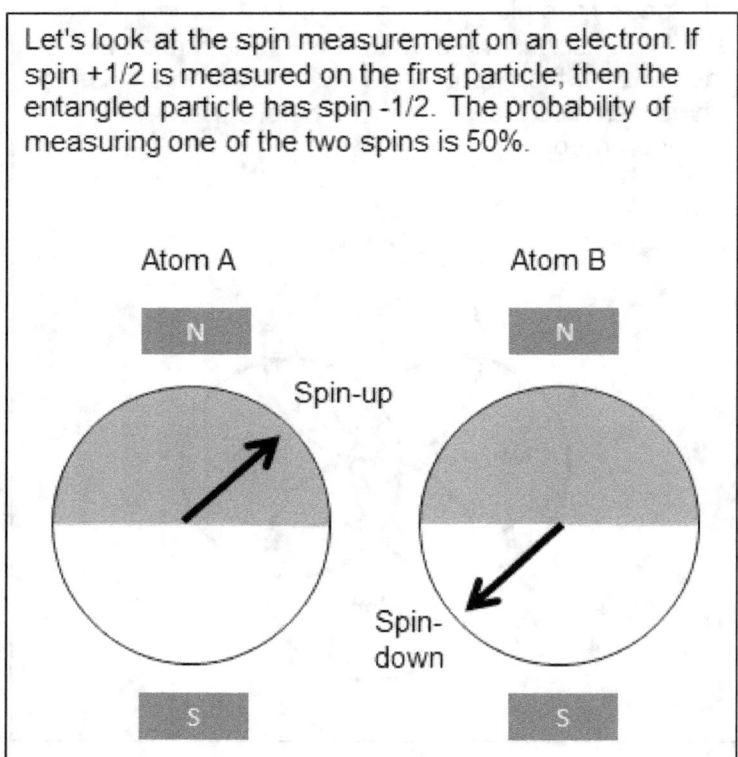

Now we slowly rotate the magnetic field, which we use to measure the orientation of the spin of the second atom. With this, the region that we register as spin-up of the second atom slowly shifts into the region that we previously registered as spin-down. The probability for the spin-up/spin-down event becomes progressively smaller and when we have rotated the magnetic field by 180° the probability is zero since we now register both events as spin-up.

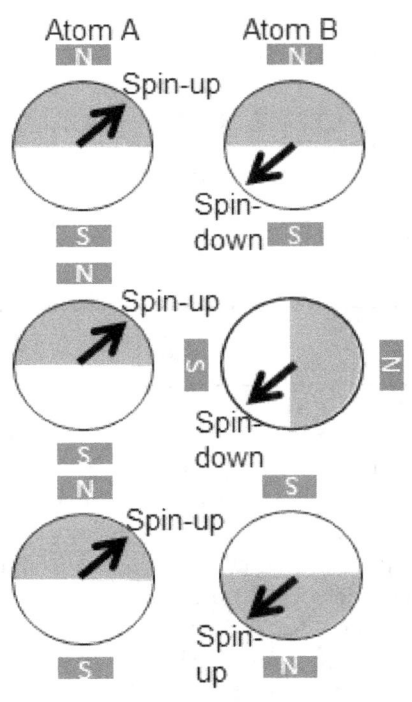

In the case of classical physics, one would expect the probability to change linearly. However, in the case of quantum mechanics, the probability is proportional to the square of the cosine. The measurement can therefore be used to distinguish whether the probability in the quantum world deviates from what is to be expected in classical physics – and whether there can therefore be no hidden parameters.

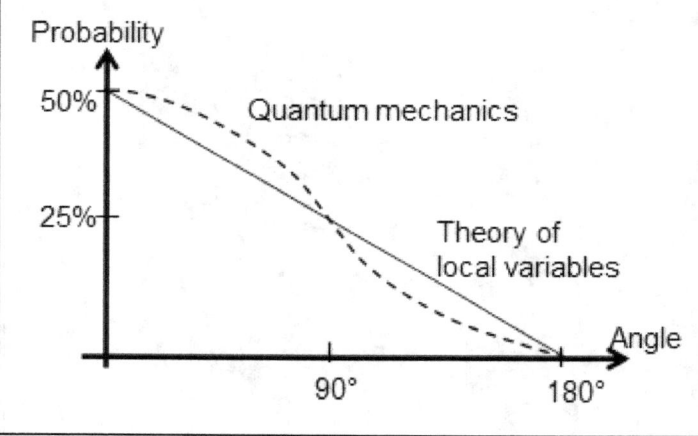

Bell's inequality can therefore be used to check whether Einstein's claim that quantum mechanics must be realistic (i.e. there are hidden parameters that describe the behavior of the particle even before the measurement) and local is actually correct. If measurements show that Bell's inequality is violated, then quantum mechanics behaves differently than Einstein believed.

Bell set up his inequality for the spin of the electrons. Entangled electrons are not that easy to produce, which is why Bell's proposal was not immediately tested experimentally.

However, in 1969, Clauser, Horne, Shimony, and Holt (Clauser 1969) described a version of Bell's inequality that used the polarization of photons. They also proposed an experiment to check the inequality. This experiment was then performed by Freedman and Clauser in 1972 (Freedman 1972). It showed that Bell's inequality is violated, i.e. that there are no hidden parameters in the quantum world.

Two entangled photons were used in the measurement and their polarizations were measured. However, prior to the measurement, it was determined which polarization should be measured. Thus, it could not be ruled out that the structure of the measurement somehow influenced the result. This influence could only be ruled out, when it was possible to determine which polarization should be measured after the photons had been sent on their journey. Such an experiment was carried out by the French physicist Alain Aspect in 1982 (Aspect 1982). This experiment again showed that Bell's inequality was violated. It is considered to be *the* experiment that clearly showed for the first time that quantum mechanics behaves differently from classical physics.

Aspect also used entangled photons generated in a special source in his experiment. This can be done, for example, by converting a high-energy photon into two photons of lower energy using a barium borate crystal. Due to their joint formation process, they are entangled with each other, and the polarization directions of the two photons are perpendicular to each other. Now the photons are emitted in different directions and sent through polarization filters. These are rotated by certain angles, and one measures how the polarization of the two photons correlates with each other. However, thanks to a very fast setup, the angle of the measurement was set after the photons had been sent on their way. Again, it was shown that the correlation is not what one would expect according to Bell's inequality if there are hidden parameters, but that Bell's inequality is violated.

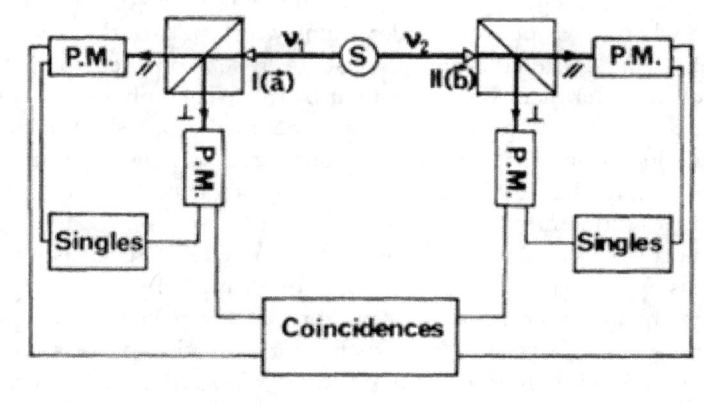

The set-up of the experiment by Alain Aspect. A source S delivers entangled photons. Polarimeters I and II transmit parallel polarized light and reflect perpendicularly polarized light downwards. The polarimeters are rotatable and are adjusted after the photons have left the source. Then the orientation of the photons is measured.

However, there were still small doubts about Aspect's result, since he could not measure all photons with his setup. Many were simply lost in the apparatus. Were the measured photons then representative of all photons, or would we have seen that Bell's inequality is not violated if all photons had been measured?

But the experiments became more sophisticated and accurate, and in 2015 researchers from the US and Canada published results that should leave no doubt (Giustina 2015).

For quantum mechanics, therefore, the probabilities are different from those of "classical" physics. There don't seem to be any hidden parameters. This can be explained in two ways: either the quantum world is not realistic or it is non-local. A non-realistic quantum world would mean that physical objects are only defined when they are measured, but do not really exist before that. Even if this is the credo of the Copenhagen interpretation, many physicists today have problems with this interpretation. Ultimately, this would mean that the world is just a creation of our minds, as Wigner's friend paradox shows – and then there could be no universally applicable laws. A non-realistic world could also be over-realistic: Not only are the parameters of the quanta fixed before the measurement, but how we measure has also been fixed for ages. We have the impression that we would set the measurement conditions shortly before the measurement, but in reality, they have been defined by the physical rules since the beginning of the universe. Then we would also get results that appear to be random but are not random at all. Since we believe that physical rules determine everything, we cannot rule out this possibility for the time being. However, complex systems behave chaotically – and the universe and humans are certainly complex systems – so that their behavior cannot be precisely defined over millions of years.

But if we can exclude missing realism of the quantum world, then there is only non-locality which remains to explain the violations of Bell's inequality observed in the experiment.

But his non-locality contradicts the theory of relativity. Nothing can propagate faster than the speed of light. But does this really hold in any case?

Physicists tend at the moment to assume that the speed of light is only an upper limit for material objects. This means energy, including a radio signal or any other signal, cannot propagate faster than the speed of light, since energy and mass are equivalent. But there are obviously features which do not belong to the "hard core" of an object, which are not determined by its matter or its energy content.

Let's take the example of light. Alain Aspect carried out his verification of Bell's inequality using polarized light. Light can be polarized in four ways: vertically (|), horizontally (-), left-diagonally (\), or right-diagonally (/). A transmitter, let's call her Alice, now sends a right-diagonal signal. However, the recipient, Bob, does not know which signal Alice has sent. He uses a filter that allows him to clearly distinguish vertical and horizontal polarizations (for example, vertically polarized light could have the maximum intensity, horizontally polarized light would not get through). The diagonally polarized light now hits this filter. It contains both horizontally and vertically polarized components. Therefore, Bob sometimes measures a horizontal and sometimes a vertical signal. He cannot be sure whether the original signal was polarized horizontally, vertically or somehow diagonally, since he can only detect the polarization with a certain probability. Thus, he needs to communicate with Alice to find out what the original polarization looked like. However, this communication certainly takes place at a speed that does not exceed the speed of light.

Quantum mechanics allows the interaction of quanta with a velocity that is above the speed of light, the so-called action at a distance; it is non-local. However, the measurement results are known only with a certain probability. Thus, no information can be exchanged. The signal velocity, the speed at which information or energy is exchanged, is not greater than the speed of light in quantum mechanics as well. The non-locality of quantum mechanics, which concerns properties that can only be measured with a certain probability, i.e. which do not allow a signal to be transmitted faster than the speed of light, would thus not contradict the theory of relativity.

*

And in fact, applications for the non-locality of the quantum world have already been found today.

One of these applications is the phenomenon of quantum teleportation. Hearing "teleportation" one thinks of beaming known from Star Trek, but unlike beaming no objects are brought from one place to another, but the properties of a particle are brought to another particle.

The existence of this phenomenon was theoretically predicted in 1993 (Bennett 1993). Quantum teleportation was first realized in 1997 by Dik Bouwmeester and colleagues in Anton Zeilinger's Vienna group (Bouwmeester, 1997).

To transfer the property from one particle to another, one takes two particles that are entangled with each other, particle a and particle b. Particle b should get the quantum state of a particle, which we will call c. For this to happen, a is entangled with c in a so-called "Bell measurement". As a is now entangled with c, and a was entangled with b, this means that b has now the same properties as c (as a is identical). But due to entanglement alone it is not yet possible to say what the entanglement really looks like. Mathematically, there can be four possibilities of how the states (e.g. spin-up and spin-down) of the two particles relate to each other. These states are referred to as "Bell states". They are written as follows:

$$State\ 1 = \frac{1}{\sqrt{2}}\left(|0\rangle_a|1\rangle_c - |1\rangle_a|0\rangle_c\right)$$

$$State\ 2 = \frac{1}{\sqrt{2}}\left(|0\rangle_a|1\rangle_c + |1\rangle_a|0\rangle_c\right)$$

$$State\ 3 = \frac{1}{\sqrt{2}}\left(|0\rangle_a|0\rangle_c - |1\rangle_a|1\rangle_c\right)$$

$$State\ 4 = \frac{1}{\sqrt{2}}\left(|0\rangle_a|0\rangle_c + |1\rangle_a|1\rangle_c\right)$$

Only after the Bell measurement has been performed, the information, in which state the entangled system is, is really known, and thus the state of b is known for sure.

The transfer of the state happens through the entanglement, it is non-local, but afflicted with an uncertainty, since theoretically there could be four states. And the communication, which one of them is realized, is done locally. Thus, Einstein's theory of relativity is not violated.

To transfer the property from one particle to another, one takes two particles, a and b, which are entangled with each other. Particle b should get the quantum state of particle c. To achieve this, a is entangled with c. Since a was entangled with b, c is also entangled with b. Its properties have been teleported instantly. But there are four possibilities how the entanglement can be realized. Only after measuring the entangled system a and c, we know in which of these states the particles really are. And this information will be passed on locally.

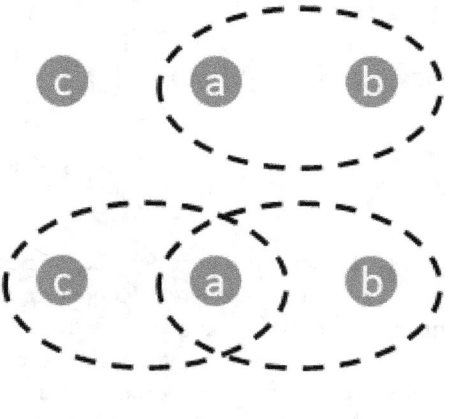

*

Entanglement is also used in quantum cryptography. The term quantum cryptography is somewhat misleading: it is not about encrypting a message using quantum mechanics. Quantum cryptography also uses an encryption with classical methods. But with these classical methods, a key must be exchanged between the two partners, between Alice and Bob. Of course, this handover of the key can be intercepted, which makes the whole encryption worthless, because now the eavesdropper, known as Eve (after the word "eavesdropping"), can decrypt the entire message.

To prevent this, one uses quantum mechanics. One takes again entangled particle pairs, and Alice sends one of the particles to Bob. If this particle is now intercepted by Eve, then she will change it by her measurement. Bob can detect this change with high probability. So quantum cryptography does not ensure that the key can be exchanged securely between Alice and Bob, but it only ensures that it is possible to detect whether someone has intercepted this key; because if Eve is listening in, then she will have measured the correct value in only one of four cases (there are, after all, four Bell states), and will pass this on. So Bob can quickly tell if someone has hacked into the transmission of the key, since it now has a lot of errors.

This quantum cryptographic encryption is already in use. The very first money transfer with quantum cryptographic encryption took place on April 21, 2004 in Vienna. The researchers from the University of Vienna in Professor Anton Zeilinger's working group had laid a 1,500-meter-long optic fiber cable through the city's sewers from the town hall to Bank Austria. The described procedure was then used to transfer a donation of 3,000 euros from the City of Vienna to the University of Vienna. Today, this information can be exchanged over hundreds of kilometers.

*

Another application of non-local entanglement is in quantum computing.

Quantum computers are the great hope of engineers because they can significantly speed up calculations. Two features of the quantum world are decisive for this: Superposition and entanglement.

Conventional computers only work with one bit: A status can either have the value 0 or the value 1. Quanta, on the other hand, can also assume values in between, since their state is unclear. According to the Copenhagen interpretation they can have any state in between because before the measurement they are in a superposition of all these states, according to the de Broglie-/Bohm-interpretation they can be in any state because we cannot determine exactly in which state they are. This means that a bit of a quantum computer can assume more than two states, theoretically an infinite number. One therefore also speaks of a quantum bit or qubit.

But the real highlight is the entanglement. If one has several quanta, which are entangled with each other, and performs an arithmetic operation on one quantum, then this influences the other quanta instantaneously. Calculations can thus run infinitely fast – if one would not need a finite time to manipulate a quantum. Thus, with a few thousand quanta, one can probably perform calculations that would otherwise require a supercomputer. However, it is tricky to create entanglement between thousands of quanta. At the moment, this can only be done for a few hundred, and this entanglement does not last very long. But the development of quantum computers is still in its infancy. There is hope that in the medium-term quantum computers will be superior to classical computers.

Thus, non-local entanglement is already being used – without being understood exactly how it actually works.

7. What we know so far

At this point, let's recap what we can actually say with certainty about the quantum world – and really about the quantum world and not a quantum world as seen by the Copenhagen or any other interpretation.

We know that there are two phenomena in the quantum world: one local and one non-local, one phenomenon traveling no faster than the speed of light, and one phenomenon (or phenomena) traveling faster than the speed of light (see also Aharonov 2017).

We know the first from the measurements in the quantum world: We can only measure a particle, even in experiments in which we examine the wave nature of the quantum world, such as the interference at the double slit. Even in this case, we ultimately only measure a particle. And that particle, be it a photon or an electron, does not move faster than with the speed of light. The photon has no rest mass and therefore moves at exactly the speed of light, the electron has rest mass and therefore moves at speeds below the speed of light. And indeed, experiments in which the electron is accelerated show that the mass of the electron increases as it was predicted by Einstein in his theory of relativity. So, we definitely have a phenomenon in the quantum world that can be identified with the particle, and it behaves locally.

Now, however, the measurements to Bell's inequality have shown that there are also phenomena which behave non-locally. Remarkably, these are phenomena that we would understand as properties of particles, such as the polarization and the spin of a quanta. However, while particles cannot travel faster than the speed of light, their properties propagate at faster-than-light speeds; they are non-local.

Einstein originally described his paradox, on which Bell's inequality is based, for kinematic parameters such as location and momentum. The delayed choice experiment suggests that these can also change non-locally.

The experiments show that a quantum has two properties: on the one hand, a local particle, which is also measured in the double-slit experiment, and on the other hand, properties such as spin and polarization, which do not propagate locally, as measurements on Bell's inequality have shown.

Local
particle

Non-local properties

We thus have the picture that particles move at most at the speed of light, but their properties as well as their kinetic parameters can change faster than the speed of light. The de Broglie-/Bohm-interpretation envisages exactly this (which is why it was initially rejected, since nothing was allowed to move faster than the speed of light): The particles follow a path that is defined by a guiding wave or a quantum potential, with the guiding wave or the quantum potential changing faster than light.

Bohr had claimed that quantum mechanics is non-deterministic. It turns out that it is deterministic after all. However, Einstein insisted that it must be local, but it turns out that quantum mechanics shows non-local effects. Ultimately, both had correctly recognized that the other's assumptions could not be correct to explain the quantum world, but they had not recognized that some of their own assumptions could not be correct either. When asked who was right in the great discussion about the interpretation of quantum mechanics, we come to the conclusion that both were partly right – and partly wrong.

The particle and the wave character of a quantum differ fundamentally: The particle behaves locally, the wave non-locally. The two descriptions are not complementary, as assumed in the Copenhagen interpretation. Particle and wave are not equivalent descriptions of a quantum, and one chooses sometimes one and sometimes the other depending on the situation, but they are the two descriptions needed to fully describe a quantum.

That particle and wave are not equivalent descriptions for a quantum, but describe different parts of the quantum, is already known. We know that a photon travels at the speed of light, c. The velocity is given by the product of the wavelength λ with the frequency ν. The wavelength describes the wave behavior of the particle, and the frequency (multiplied by Planck's quantum of action) provides the energy of the particle. The formula

$$c = \lambda \nu$$

unites the wave and the particle image. Now we let a light wave shine through glass. In matter, light travels slower than in air, by a factor described by the refractive index n. For the velocity in matter we have:

$$c_{Matter} = c/n$$

If particle and wave were equivalent descriptions of a quantum, then the frequency and the wavelength would have to change simultaneously. In fact, however, the frequency remains unchanged and only the wavelength is reduced by the factor of the refractive index. This is often explained by the fact that the energy of a photon (given by the frequency) cannot change when passing through matter. But why does the wavelength change when the wave is an equivalent description of the particle, and both are just two sides of the same coin? If one changes, the other should change too.

However, this is not the case because the particle and the wave are two facets of the quantum world. The wave "tells" the particle how to behave, non-locally, the particle follows this specification, whereby it behaves locally, i.e. does not move at a speed greater than the speed of light.

If a wave passes through a medium, the speed of light c decreases by a factor of the refractive index n. The wavelength λ is also reduced. The frequency v, on the other hand, remains unchanged. Wave and frequency are not equivalent and complementary descriptions of a quantum.

c	c/n
λ	λ/n
v	v

The properties of a particle and its behavior are not determined by the particle directly, but by the associated wave. A particle itself cannot be polarized, for this it needs a wave; and with this wave the information of the polarization spreads non-locally. We will deal later with other properties which are not part of the particle, but for the moment we want to leave the world of the quanta and turn to the second great theory of the 20th century with which the quantum theory should be unified: The theory of relativity.

8. The theory of relativity

8.1. The special theory of relativity

What actually happens when you move faster and faster so that you eventually move with a light wave? If you move as fast as the light – does the wave then still move or does it seem to stand still? Albert Einstein dealt with this question at the beginning of the 20th century.

Behind this thought experiment of Einstein's is the question of how physics changes from one moving system to another. The answer that has been given all these years, and that Einstein should also give, is: Not at all.

Think of a person in a moving train and a person being on the platform. If the person on the platform throws a ball to the ground, then the ball bounces back from the ground to the person. The same happens in a moving train. A human being cannot distinguish from this experiment whether he is standing on a platform or in a moving train. The physics is the same for both.

If both people now go in one direction, then they will move relative to the ground, for example, at a speed of 4 km/h. Here, too, neither of them can tell whether they are in a moving train or on the platform. This is different when the person on the platform observes the movement of the person on the train. If the train is moving at 100 km/h and the person on the train is moving at 4 km/h towards the end of the train, then the person on the platform will notice that the human on the train is moving at 96 km/h away from the stationary person on the platform. The speed of walking is subtracted from the speed of the train, since both are moving in different directions. This is known as Galilean transformations. They allow to convert the frame of reference of people on the platform into the frame of reference of people on the train; the corresponding speeds are simply added or subtracted.

If you apply this procedure to the light wave and to someone who moves with the light wave at the speed of light, then you get a paradoxical result: After the Galilean transformation, the difference in speed between the light wave and the observer is zero. For the observer, the light wave would then stand still.

If a car moves forward at 100 km/h and a passenger moves to the end of the car at 4 km/h, then an outside observer will calculate a speed of 96 km/h for the passenger according to the Galilean transformations. If the outside observer is also moving at 100 km/h, then the car appears to be standing still and the observer is moving away at 4 km/h.

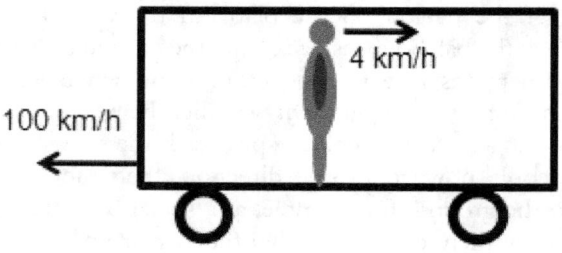

A light wave moves with the speed of light c. For an outside observer also moving with c, the light wave would stand still – but then it would cease to exist.

On the other hand, the propagation of a light wave is explained in such a way that the electric fields of the electromagnetic waves produce magnetic fields, which then in turn produce electric fields, and so on. In order for a wave to exist at all, it has to move. A standing wave could not generate fields – and would cease to exist. This means that for a stationary observer the light wave would exist, for someone moving at the speed of light it would not exist. Both would experience different physics. That was ruled out in classical physics, and Einstein also considered it impossible. Something couldn't be right here.

A similar problem existed with the Maxwell equations. At the end of the 19th century, the physicist James Clerk Maxwell set up four small equations that describe electromagnetic phenomena such as the emission of electromagnetic radiation, induction or the shape of electric and magnetic fields.

The wave equation, derived from the Maxwell equations, contains the speed of light. And that speed of light is a constant. No matter what reference system you are in and how fast it is moving, the wave equation assumes that the speed of light is always constant.

The Maxwell equations are:

$$\nabla E = \frac{\rho}{\varepsilon_0}$$

$$\nabla B = 0$$

$$\nabla \times E = -\frac{\partial B}{\partial t}$$

$$\nabla \times E = \mu_0 j + \mu_0 \varepsilon_0 \frac{\partial E}{\partial t}$$

The wave equation is:

$$\frac{1}{c^2}\frac{\partial^2 E}{\partial t^2} - \nabla^2 E = 0$$

with $c^2 = \mu_0 \varepsilon_0$ independent of a reference system.

This has the consequence that the Galilei transformations do not work anymore for the Maxwell equations. If you transform the equations from a stationary system to a moving one, then you receive expressions which describe completely different physics. But a basic assumption of physics is that the physical rules should be identical independent of the reference frame. In mechanics this still worked, in electromagnetism of the Maxwell equation this suddenly does not work anymore.

The physicist Hendrik Antoon Lorentz suggested that Maxwell equations would be valid only for a specific reference frame. This specific reference frame was the absolute space in which the ether rested, a substance in which the electromagnetic waves propagated. In this ether the speed of light was constant, if one moved relative to the ether, then the speed of light would change. According to Lorentz, if light moves in the same direction as the source moves, then its speed should be different from that perpendicular to it. The problem is that the speed of light is very high. It is 299,792,458 m/s, i.e. almost 300,000 km/s. A car moving at 200 km/h only covers about 55 meters in one second – that's nothing compared to the speed of light. The earth moves around the sun at almost 30 km/s, i.e. one ten-thousandth of the speed of light. The sun – and thus the earth – moves around the center of the Milky Way at about 220 km/s, i.e. almost a thousandth of the value of the speed of light. And with that you get to an order of magnitude where you can at least measure changes in speed with the help of interference.

In 1887, the American physicists Albert Michelson and Edward Morley conducted an experiment using an interferometer. Light was split into two beams: one beam led away from the light source, was reflected and then directed onto a screen. The second beam moved back and forth perpendicular to it before being directed onto the same screen. As with the double slit, the two beams created an interference pattern on the screen.

If one turned the interferometer, then one branch of the interferometer, which had pointed so far in the direction of the earth's movement, would now be perpendicular to it, and the other would be pointing in the direction of the earth's movement. The velocities of the light relative to the ether should change in the two branches of the interferometer. Since this also changes the distribution of wave crests and troughs, the interference pattern should change. Thus, it was possible to measure small changes in the speed of light in the Earth's reference frame. However, Michelson and Morley could not measure any change of the speed of light. It was constant in all directions, no matter how the earth moved to the ether.

In the Michelson interferometer, the light is split by a semi-transparent mirror, directed onto two mirrors and then collected on a screen, where it shows an interference pattern.
The pattern does not change no matter how you point the interferometer to the direction of Earth's motion.

This result was unexpected. In 1895, Hendrik Lorentz tried to explain this result by saying that the branches of the interferometer were compressed or lengthened, depending on how they were positioned relative to the ether. With his transformations he was able to explain the measurement of Michelson and Morley and surprisingly they were also compatible with the Maxwell equations, which was not the case with the Galilean transformations. But how should the ether compress physical objects? For that he would have to provide a lot of power, but you couldn't even feel the ether. This explanation was therefore not really satisfactory.

In 1905, Albert Einstein chose a new approach (Einstein 1905a): If we measure that the speed of light is constant for all frames of reference, then that must be the starting point for our considerations. Therefore, he postulated the constancy of the speed of light in all frames of reference and then thought about what the transformation rules for frames of reference that move at different speeds would have to look like. In doing so, he came up with the same transformation rules that Lorentz had already derived under completely different assumptions.

Two notable effects result from these rules: time dilatation and length contraction.

Let's take a simple example: observer A is sitting in a moving train; observer B is standing on a platform. Light moves from the ceiling of the train to the floor. For the moving observer A, the light needs the short time of 0.01 μs for the distance of 3 meters. The situation is somewhat different for the stationary observer B. The train moved in a time of 0.01 μs. Let's say it moves a meter (it's an unusually fast train). Then the light moves for him on a diagonal of the triangle, which is formed by the height of the train and the moved distance. According to Pythagoras, the distance traveled by the light is about 3.16 meters; it has to travel a longer distance.

According to the Galilean transformations, the speed of the train and the speed of the light add up and time remains constant. According to Einstein, however, light moves at the same speed in both reference systems. Since it has to travel a longer distance for the unmoving observer, it follows that time passes a little more slowly for someone riding the train; because to cover a distance of 3.16 meters, the light needs 0.0105 μs. This is the so-called time dilatation: A stationary observer B must assume that the time passes more slowly for a moving observer A, although the latter, when measuring the time himself, also believes that only 0.01 μs have passed.

For a moving observer, the unit of time is determined by the time it takes for a light signal to be reflected by both mirrors. The time unit is equal to 2d/c.

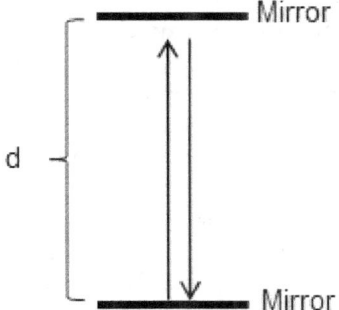

For a resting outside observer, the light in the moving system covers a longer path. Since the speed of light is constant, it takes a longer time to cover the way. We have a time dilatation.

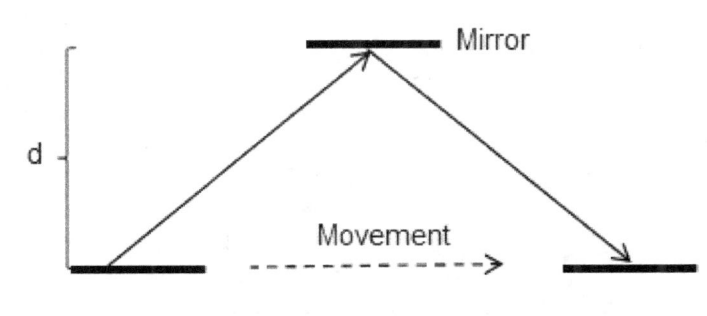

This time dilatation is accompanied by a length contraction. Since for an observer B at the platform the time seems to pass faster than for the observer A in the train, he measures a shorter length for a moving measuring unit using a light ray. The length for the measuring unit seems to be contracted (and both time dilatation and length contraction are true for the observer A as well, as no one can tell, which observer is moving).

Time dilatation is accompanied by a length contraction. The bottom line is that the same physical laws thus apply in both systems, there are no more contradictions.

The existence of time dilatation and length contraction cannot only be justified theoretically, but also verified experimentally, namely based on the lifetime of a muon, a "heavy" electron. It is created by cosmic rays in the upper atmosphere, at an altitude of about 10 km. The muon at rest has a lifetime of only about 1.5 μs, as has been experimentally proven. This means that the muons created in the upper atmosphere would only travel 500 meters at a speed close to the speed of light. Nevertheless, they can also be detected on the surface of the earth. In fact, at this high speed, muons have a lifespan of about 13 microseconds, and some live as long as 30 microseconds. They live much longer for an outside observer, as predicted by the theory of relativity – and easily reach the earth's surface, where they can be measured. In 30 μs, muons can easily cover 10 km.

At the same time, the muon experiences a length contraction. Instead of a distance of 10 km, it experiences only a distance of 500 m, which means that it also reaches the ground with a lifetime of 1.5 μs. Time dilatation and length contraction are two sides of the same coin. Space and time change simultaneously for a moving observer. Since this is the case, Hermann Minkowski introduced the concept of space-time in 1907 (Minkowskis 1907): Space and time are connected. If space changes in one direction, time must change in the other direction. Our world can thus be understood as a four-dimensional world in which a line element remains constant independent of the observer. This line element is a distance in the four-dimensional space-time. The length of a line is calculated according to Pythagoras by calculating the square of each coordinate and then taking the square root of the sum of the coordinates. This line element is an invariant of space-time: No matter from which reference frame you observe it, you always get the same value; because in space-time, space and time behave in opposite directions. Therefore, the three place coordinates get a positive sign, the time coordinate a negative one, whereby the time coordinate is also multiplied by the speed of light to get the unit of a distance.

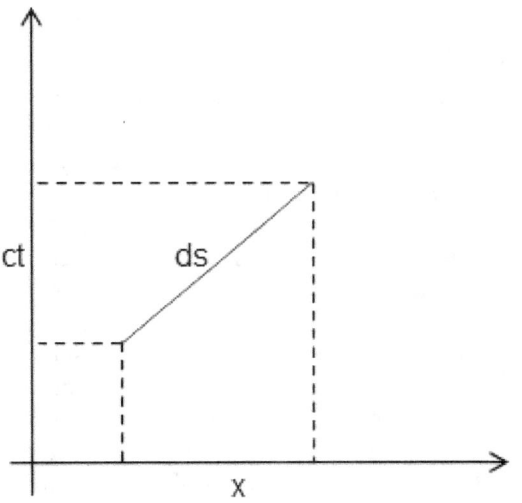

A line element (shown in the figure for only one space dimension) in spacetime is defined as

$$ds = \sqrt{x_1^2 + x_2^2 + x_3^2 - (ct)^2}$$

The line element is a constant. If the proportion of space increases due to relativistic effects, then the proportion of time is reduced accordingly.

In the same year that Einstein published the special theory of relativity, he published yet another article (Einstein 1905b) in which he derived his famous formula $E=mc^2$ from the formulas of the theory of relativity. This formula says that mass and energy are equivalent, and a small mass contains a lot of energy because it is multiplied by the speed of light squared. So, there is a lot of energy in a small mass, as can be observed in the explosion of an atomic bomb.

In the special theory of relativity of 1905, Einstein was concerned with how space and time behave for observers moving at a constant speed relative to one another. But how do time and space change when the observers are accelerated relative to each other?

8.2. The general theory of relativity

To understand how space and time change when a frame of reference is accelerated, Einstein thought of a simple example: He looked at a rotating disk. An observer on the edge of the disk experiences an acceleration because he is constantly changing his direction. However, the disk should rotate at a constant speed, so that the direction of the speed changes, but not its amount. Then Einstein drew tangents to the circumference of the circular disk in his mind. These tangents are approximations of circular motion. If you choose them small enough, then you can use these straight lines to approximate a circular movement as precisely as you like. However, there is no acceleration on these small tangent sections; the observer traveling along only experiences an acceleration when he changes his direction slightly at the next tangent section. According to special relativity, length along the circumference appears contracted because each tangent segment points in the direction of the movement. This is different for the radius: The radius is perpendicular to the direction of the movement and does not change its length. For a disk at rest, the circumference is twice the radius times the number π, i.e. radius times 2π. If the circumference decreases and the radius stays the same, then this rule no longer applies. With accelerated reference systems, the well-known geometry of everyday life, which is also known as

Euclidean geometry after the Greek mathematician Euclid, no longer applies; the space appears to be deformed.

> The edge of a rotating disc can be divided into many small tangent pieces. In a first approximation, these do not experience any acceleration, but move at a constant speed, where they experience a contraction in length.
> The radius, on the other hand, is perpendicular to the movement and does not experience any change in length. This means that the perimeter is no longer equal to the radius times 2π. In an accelerated system, the well-known Euclidean geometry no longer applies.

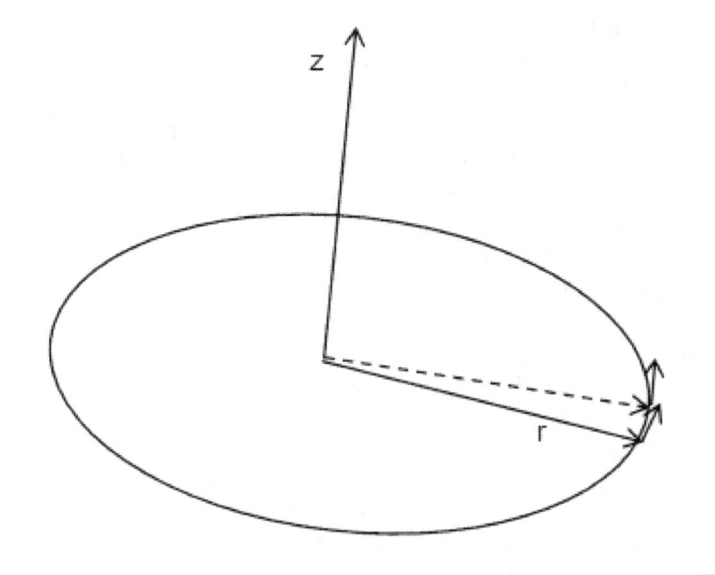

This was already indicated by the Minkowski representation of space-time: space and time are interwoven in such a way that time influences space. This means that in all generality the Euclidean geometry cannot be valid in space-time.

A general theory of relativity that also takes acceleration into account would not only explain how space and time are converted in mutually accelerated reference frames. As Einstein realized, such a general theory of relativity would also explain gravity. He came to this conclusion because, as in the case of special relativity, he made a simple assumption.

When Newton wrote down his three laws, he said that force is proportional to acceleration. He called this proportionality factor mass. Strictly speaking, this is the inertial mass, since it opposes a change in speed, an acceleration. At the same time, Newton established a law of how bodies behave in a gravitational field, the law of gravitation. A mass also appears in this law, the gravitational mass. First of all, nothing suggests that both masses have to be identical, after all we experience one mass during an acceleration, the other in a gravitational field, i.e. under different physical conditions. But physically one couldn't see any difference, so one gave up the distinction and just talked about one mass without really understanding why they should be identical.

Einstein was the first who could explain this: He imagined an elevator in space. In this elevator there is a human being who is, of course, weightless in space. If one brings the elevator into the gravity field of a planet, then elevator and man are attracted, and the man moves to the ground of the elevator. He feels his gravitational mass. If you leave the elevator in space and set it in motion, then the elevator pulls the human with it. The person stands firmly on the floor of the elevator and is accelerated by it. The acceleration also leads to the fact that the human being no longer glides weightlessly through the elevator, but has the impression to be attracted by the ground that approaches him. Ultimately, man cannot distinguish acceleration from gravity. They are only two sides of the same coin, which is why it was justified to consider gravitational and inertial mass to be equal. But with that a theory of relativity for accelerated motion would explain gravity at the same time: Just as accelerations change space-time so that it is no longer Euclidean, gravity would also change space-

104

time so that it is no longer Euclidean. After all, both are just two different views of one physical phenomenon.

An observer in an elevator cannot distinguish whether he is being pulled towards the bottom of the elevator by a gravitational field or whether he is observing this due to an acceleration. Acceleration and gravity are equivalent, gravitational and inertial mass must be equal.

$F=mg$

$F=mg$

Objects generally move in space-time along a line called a geodesic. This means the shortest connection between two points. In Euclidean space, this is simply a segment given by a straight line. Objects are not accelerated on this geodesic. This is different in curved space-time: Here, the movement along a geodesic leads to an acceleration of the objects.

On the earth's surface, the geodesic is given by a great circle running along the surface of the earth. The big question was: What would the geodesic look like on which bodies generally move in space?

Bodies move along the line of shortest distance between two points, on a geodesic. On a sphere like Earth, this is a great circle. The sum of the angles of a triangle on a sphere is greater than 180°, i.e. this surface is not an Euclidean surface.

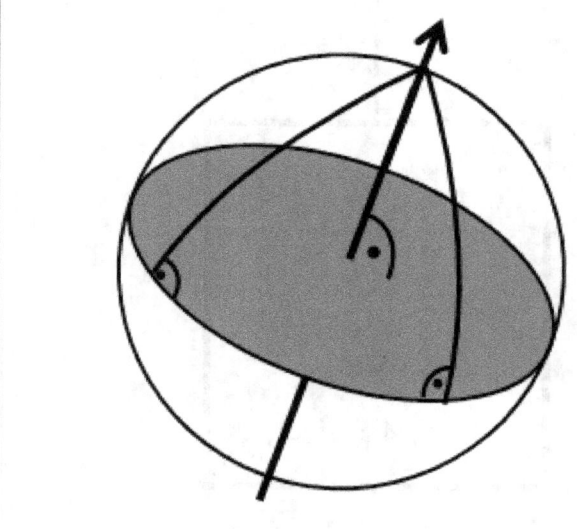

Einstein knew that gravity altered space-time. All he had to do was to find a formula that linked mass or energy to the curvature of space-time. However, it took him ten years to do this, since he first had to deal with non-Euclidean geometry, which even today is only taught in advanced courses at the university.

This geometry goes back to the German mathematician Bernhard Riemann, who lived in the 19th century. Their mathematical formulas allow geometries to be described in any number of dimensions with any metric. The metric describes the local coordinate system, which can differ from place to place. In the special theory of relativity, it is given by Minkowski's representation and consists of a flat space-time without masses. The space of this geometry is Euclidean.

To describe general deformations of space, Riemannian geometry uses so-called tensors. Tensors map objects to each other. A zero-order tensor is a scalar, i.e. a number. A simple mapping consists of multiplying a number by a scalar (i.e. a number) to get another number. A tensor of the first order is a vector, a tensor of the second order is a matrix, a tensor of the third order is then already difficult to fit on a page, it would practically consist of several matrices arranged one behind the other. Even if it is difficult to write down tensors, they allow to describe mathematically mappings in any number of dimensions. That's why they are used so extensively in the Riemannian geometry.

The field equation of the general theory of relativity, the curvature of space is described on the left, the curving mass/energy is on the right:

$$R_{\mu\nu} - \frac{R}{2} g_{\mu\nu} = \frac{8\pi G}{c^4} T_{\mu\nu}$$

with:
$R_{\mu\nu}$ the Ricci curvature tensor
R the Ricci curvature scalar
$g_{\mu\nu}$ the metric tensor
G the gravitational constant
c the speed of light
$T_{\mu\nu}$ the energy-momentum tensor

Einstein now found a formula which describes changes, i.e. curvatures of space-time caused by the energy in it (Einstein 1916).

This field equation actually consists of 16 coupled differential equations, which are reduced to ten by symmetries. They indicate how mass and energy bend the space-time. This then tells us how masses move along the geodesics of curved space-time. The masses deform the space, and the space then dictates how the masses have to move (for the derivation of the field equation see appendix D).

In his gravitational equation, Newton had to assume that the force somehow acts instantaneously, since it was not evident that it is transmitted at a certain speed. Einstein was able to show that gravity is a curvature of space-time caused by mass and energy, which propagates at most at the speed of light. If the sun disappeared from one moment to the next, the earth would continue its orbit around the sun for another eight minutes, since light takes eight minutes from the sun to the earth, and therefore the curvature of space-time only changes after eight minutes.

Since space and time are closely interwoven in our world, curvatures of space naturally also result in changes in time. Thus, for an object near a mass, time passes more slowly for an external observer, while the observer on the object near the mass observes a "normal" passage of time. When a spacecraft encounters a black hole so massive that even light cannot escape, the spacecraft simply flies into the black hole for a member of the crew, while for an outside observer it flies progressively slower until it seems to stand still at the edge of the event horizon, i.e. the area from which light can no longer escape.

That space changes the path of objects and even of light rays can be observed especially in the universe, where there are correspondingly large masses. Because large masses focus the light of a galaxy, gravitational lensing occurs, which is why galaxies look circularly deformed or appear several times in the sky like the famous Einstein Cross.

The curvature of space can act like a lens, creating multiple images of galaxies. An example is the Einstein Cross: the gravitational lensing effect of a galaxy produces a quadruple image of a quasar lying behind it.

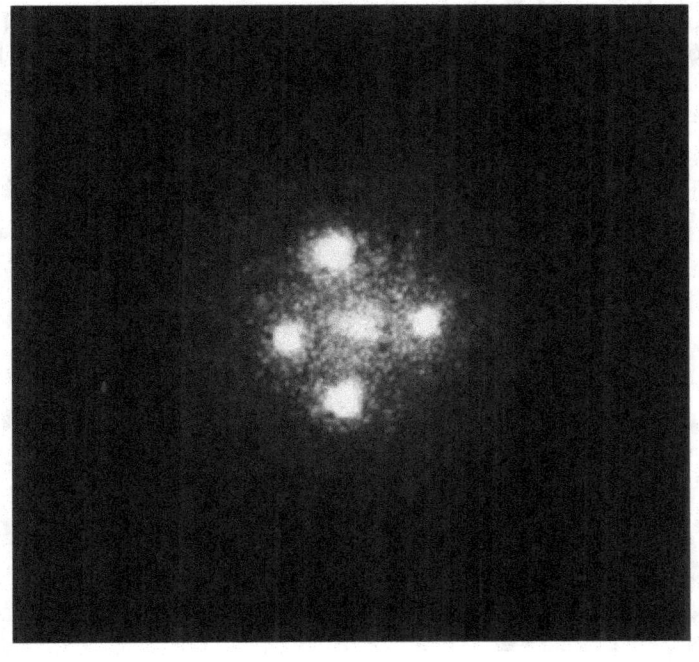

The British astronomer Arthur Eddington provided evidence of this effect of the general theory of relativity in 1919. The mass of the sun should also bend light rays. Eddington therefore measured the position of stars without a sun, then he measured the same position when the sun was in front of the stars. Since the rays of light are bent by the sun, an observer on earth would have the impression that the stars have shifted a little to the outside. Since the sun shines too brightly during the day, a solar eclipse on May

29, 1919 was used for this. And indeed, Eddington was able to announce that he had measured the curvature of the light rays, which resulted from Einstein's theory. Einstein's theory was thus considered proven, no matter how strange it might appear – and was proven again and again over the course of the next years. You can only use the GPS system with its meter-precise positioning on earth because it takes into account the time dilatation of the movement of the satellites and the time delay caused by the movement in the earth's gravitational field; otherwise, one would get wrong times and could only give the position to within a few kilometers.

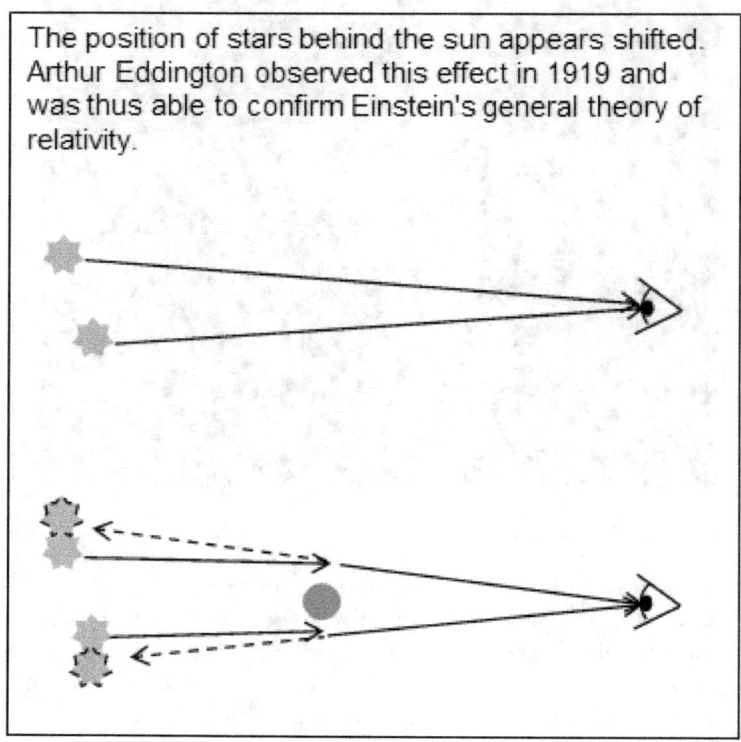

The position of stars behind the sun appears shifted. Arthur Eddington observed this effect in 1919 and was thus able to confirm Einstein's general theory of relativity.

The interpretation of gravity according to Einstein is exactly opposite to that of Newton. While Newton assumed that a force of the gravity field accelerates a body, according to Einstein the objects move force-free in a curved space. The acceleration is only a consequence of the curved space, but not a consequence of a force. On the contrary: If you are on earth, then your body wants to follow the geodesic to the center of the earth. This movement is force-free. Only when the body has reached the earth's surface, then a force acts, namely a repulsive force, which prevents the body to follow the geodesic further – and only then do we notice the "gravity".

Einstein's equations result in a dynamic universe that changes over time. When Einstein discovered this, he introduced a "cosmological constant" into his equations to keep the universe stable, since he could not imagine the universe changing. But in the 1920s, the American astronomer Edwin Hubble discovered two things: In the early 1920s, he was able to prove that the nebulae that could be seen in the sky are galaxies outside our Milky Way. Until then it was thought that the Milky Way was the entire universe, now it was recognized that the Milky Way is only a small part of the universe. Second, in the late 1920s, Hubble was able to show that all galaxies are moving away from us (except for a few in our immediate vicinity that are gravitationally bound to the Milky Way). This was reflected in a clear red shift in the light spectra of these galaxies. In addition, the further away the galaxies were from us, the greater the redshift. This can only be interpreted in such a way that space is expanding, i.e. the universe is not stable, but was created a long time ago in a big bang. Einstein therefore removed the cosmological constant from his equation and called it the "biggest foolishness of his life".

The red shift of the spectra is caused by the photons losing energy due to the expansion of space, the spectra are shifted towards the red (since red light has a lower energy than blue light). The farther away an object is from us, the faster it appears to be moving away. This can be illustrated by a balloon on which dots are painted. If you blow air into the balloon, the dots move away from each other. Dots which are already far away from each other will move away from each other by a greater distance when they are inflated, i.e. they move away faster.

Edwin Hubble observed that almost all galaxies are moving away from us. The reason is the expansion of space: This pulls the galaxies with it, just like points on a balloon move away from each other when it is inflated. The expansion of space can also take place at speeds greater than the speed of light. The speed of light limit only applies to movement in space, not movement of space.

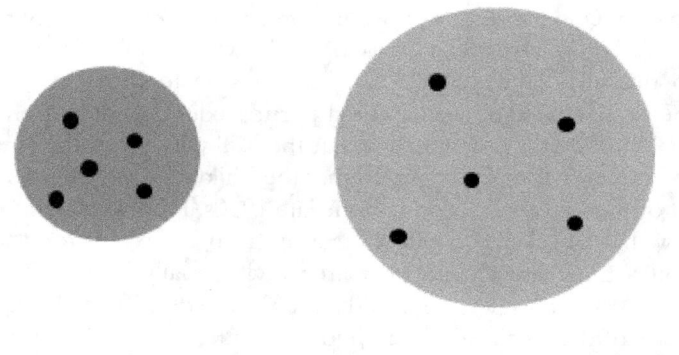

The redshift can be used to determine the speed at which galaxies are receding from us, and a surprising observation is made: there are galaxies that are moving away from us faster than the speed of light. How can this be?

Well, it's not the galaxies that are moving away from us, but space is expanding and pulling these galaxies away from us by "pulling them along". According to the theory of relativity, no object can move faster than the speed of light in space; however, the theory of relativity does not impose a speed limit on space itself. We'll come back to that later.

9. Attempts to unify quantum mechanics and the theory of relativity

The theory of relativity describes the behavior of large objects or at high speeds. Quantum mechanics describes the behavior of small objects. In the field of everyday experience, both are replaced by classical physics, which provides a sufficiently accurate description. Both theories include classical physics as a limiting case for medium-sized masses.

However, there is still no theory which describes the world from the quantum world up to the cosmos, a theory which describes the area of the quanta as well as the area of the galaxies. This "theory of everything", the great unification, is still being searched.

The theory of relativity, although it severely tests our conception of space and time and in this respect differs significantly from classical physics, is nevertheless classified as classical physics because it does not show the characteristics of the quantum world – quanta and their bizarre behavior. Since the real new thing is to be found in the quantum world, it is assumed that one has to adapt the theory of relativity to the quantum world, one has to "quantize" it. However, it is easy to forget that the bizarre behavior of the quantum world is often not a paradox of the quantum world, as we have seen, but a behavior of the Copenhagen interpretation. Therefore, one does not try to unite the quantum theory with the theory of relativity, but to adapt the theory of relativity to the Copenhagen interpretation.

One possibility for combining the quantum world with the theory of relativity is seen in string theory. This theory is currently pursued by most theoretical physicists. The building blocks of nature, the electrons and quarks, are here neither particles nor waves, but small strings. These strings oscillate with characteristic frequencies, with which they appear as different particles. The strings can be open or closed.

According to string theory, the smallest building blocks of matter are not quanta, but small strings that can be open or closed and not only form the building blocks of matter, but also transmit the forces of nature.

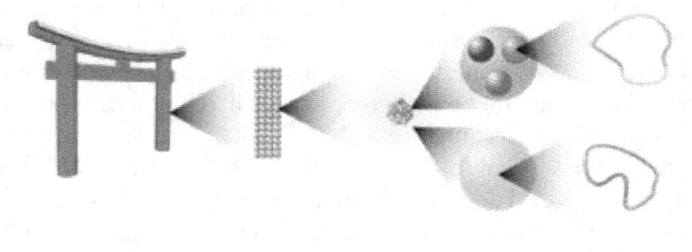

One of the reasons that string theory is seen as a way of providing a "theory of everything" is due to the fact that it involves a particle that is thought to be the fundamental particle of gravity. As Einstein showed, mass motions in space cause gravitational waves which propagate through space. These are very, very weak, nevertheless, gravitational waves were detected for the first time in 2016 – although this required two black holes to collide, and the signal was still very weak.

Since, according to the Copenhagen interpretation, everything should have both wave and particle properties, one expects that gravitational waves are also associated with a particle that is called a graviton and to which certain properties are ascribed. String theory delivers this particle free of charge. It thus has the potential to describe not only the electromagnetic force, as well as the strong and weak nuclear force, but also gravitation, which no other theory has yet succeeded in doing.

However, one has to be careful: Even if the Copenhagen interpretation assumes that a particle is associated with each wave, this is not correct in all cases. Sound waves are pure waves not connected with a particle. Sometimes „phonons" are called particles of sound, but they are in fact small excitations in a solid body which can be understood as a "quasi-particle". It is therefore not at all certain that there must exist a graviton.

The string theory is considered a hot candidate for a "theory of everything" because it not only describes the smallest particles of matter and contains three basic forces (electromagnetic, weak and strong force) but also a particle which can be identified as a graviton, the fundamental particle of gravity. String theory could therefore describe all four fundamental forces, which no theory has been able to do so far.

However, since there are gravitational waves, it is assumed that there must be a particle of gravity, as one assumes that every wave must have a particle associated with it. But not every wave has a particle character.

Moreover, the string theory has the problem that it is not a theory. It is first of all a hypothesis, i.e. an assumption that has not been scientifically tested. And it is not even one assumption, but billions of assumptions.

In fact, there are several string theories that differ in details. But all of them assume that there must be more than four dimensions. Since one does not perceive these, one claims that these dimensions are rolled up so small that one cannot perceive them. How these dimensions should be rolled up is the big question as there are billions of possibilities.

To cope with this multiplicity of string theories, one of the leading string researchers, Edward Witten, claimed in 1995 that all these string theories are nothing more than variations of a higher-dimensional theory, which he called M-theory, and which is supposed to be unique. What the M is supposed to stand for, Witten never revealed. Since strings in a higher dimension can be membranes, one assumes that M stands for membrane, but it could also stand for mystery.

A problem with string theory is that there is not just one string theory, but several. The assumption is that the string theories discussed today are all facets of a higher dimensional theory, called M-Theory.

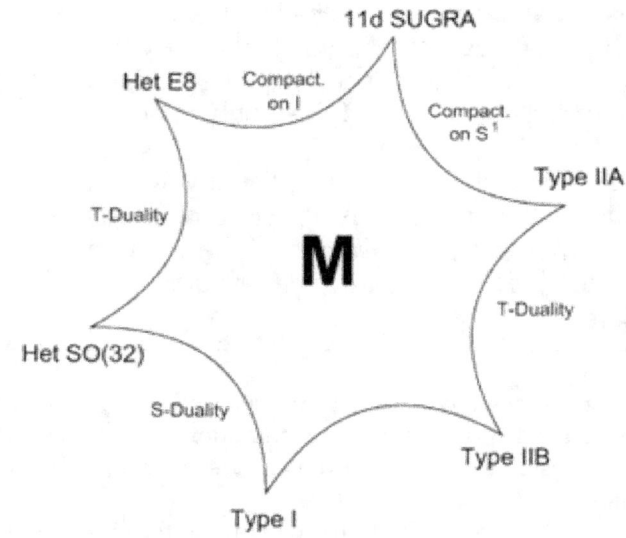

Another problem with string theory is that it is based on supersymmetry. Each fermion has a supersymmetric boson as a partner, each boson a supersymmetric fermion. Unfortunately, despite intensive searches, these super-symmetrical particles have not yet been found. The "string theory" is therefore very speculative and should better be called "string hypothesis".

In order for string theory (or string hypotheses) to work, it must assume that every elementary particle has a "supersymmetric" partner. A super partner of a quark is a squark. While a quark is a half-spin fermion, a squark is a boson with an integer spin. Unfortunately, despite an intensive search, none of these super partners have been found yet.

String theory, even if it has been praised in numerous books and films as the basis for the "theory of everything", is currently nothing more than an intellectual gimmick, albeit at the highest mathematical level (Edward Witten, for instance, received the Fields Medal for his work in the year 1990, one of the highest honors in mathematics).

According to loop quantum gravity, space consists of a network of nodes connected by lines. Properties are ascribed to the nodes that are similar to the spins of elementary particles. These nodes obey the laws of quantum mechanics and thus form space-time. The node distances correspond to the Planck length. A cubic centimeter contains 10^{99} nodes. Reactions on the quantum level change the node network.

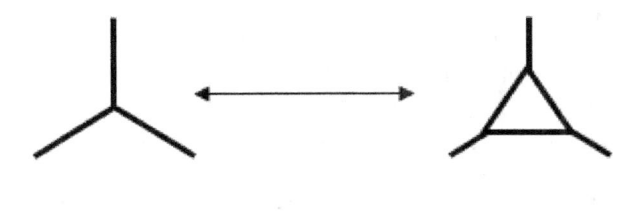

Another attempt to combine quantum mechanics and the theory of relativity is loop quantum gravity. Again, the basic idea is that one has to change space-time in order to unify quantum theory with the theory of relativity. The starting point is therefore a quantization of space-time, which, however, takes place at Planck lengths, i.e. in the range of 10^{-35} m and 10^{-43} s, i.e. again in ranges that we will probably never be able to check, such as the rolled-up space dimensions of string theory.

While string theory still assumes that there are strings in space that create the quantum world, in loop quantum gravity, the quantum world creates space. Space consists of a network of nodes connected by lines. Properties are ascribed to the nodes that are similar to the spins of elementary particles. The node distances correspond to the Planck length. A cubic centimeter contains 10^{99} nodes. These nodes obey the laws of quantum mechanics and form space-time – which is thus also quantized on its lowest level. But why do we assume that space-time must be quantized? Well, a simplified representation of quantum mechanics says that everything in the quantum world is quantized. There are particles, and the energies that an electron in an atom, for example, can emit and absorb are also quantized and cannot take on continuous values. Also other properties of the quantum world, like the spin, have only certain values and are thus quantized.

However, we have to be careful. The representation that everything is quantized is not entirely correct. There are undoubtedly particles in the quantum world that are quantized by definition. Ultimately, these particles are measured in the experiments. The fact that the energies of electrons in the atom are also quantized and cannot assume continuous values does not result from a quantization but from the boundary conditions of the wave function. The wave properties (which are inherent to the quanta, as the Copenhagen interpretation means, or guides the particles, as de Broglie/Bohm say) result in the fact that energies can only occur in quantized form. The quantum properties are a consequence of the boundary conditions (here: that the wave function spans around the atom and can therefore only exist where it is positively amplified), not the cause. Therefore, first of all, it is not clear whether everything really has to be quantized from scratch, as loop quantum gravity assumes, or whether quantization is just a consequence of another property, such as the wave nature, where a wave is not quantized. Likewise, it is also not certain that there must be a graviton, because a particle does not have to be connected to every wave, as is sometimes assumed for simplification.

Both theories have to struggle with difficulties which ensure that they are still far from being able to describe a "theory of everything". The reason may also lie in the fact that both theories do not contain the two features which we have identified as fundamental for the quantum world: On the one hand, the local particle character of quanta, and on the other hand, the non-local entanglement of the properties of quanta.

The reason for this absence is that most quantum theories do not make this distinction either, an exception being the de Broglie-/Bohm-theory, which, however, occupies rather a niche position in current research. Only a quantum theory which takes these physical facts into account, however, would have the potential to provide a "theory of everything".

10. On space

A quantum consists of a particle that moves locally and something non-local that connects entangled particles together. This non-local part includes properties such as spin and polarization, but it also affects a particle's motion. At first it sounds as if this statement would contradict the theory of relativity. However, this is not the case. The theory of relativity only forbids that things can move non-locally in space (i.e. faster than the speed of light), but space itself can move at any speed. If we now assume that the non-local properties of a quantum are properties of space (not properties in space), then we have a bridge that can be built between quantum theory and the theory of relativity.

In order to be able to put both theories together, one has to understand which tasks space takes on in the quantum world. The experiments on Bell's inequality, like the experiments by Aspect, or on quantum teleportation, have already shown that there are properties in the quantum world that are not bound to the particle, but to space through entanglement. And there are other experiments that show this.

10.1. Aharonov-Bohm Effect

In 1959, Yakir Aharonov and David Bohm published an article predicting an effect where there should be no effect (Aharonov 1959). The starting point was the setup of the double-slit experiment. Only this time, a coil is placed behind the double slit and between the two slits, creating a magnetic field, but only inside the coil. First one lets electrons fly through the double slit with the coil switched off and finds the usual interference pattern. Then one switches on the coil. This coil generates a magnetic field, but by its construction only inside the coil, not outside. The electrons flying past the outside of the coil through the double slit can therefore not feel the magnetic field. Nevertheless, the interference pattern, which can be seen on the screen behind the double slit, changes as if the magnetic field had influenced the path of the electrons. This effect was then also experimentally shown in 1986 (Tonomura 1986).

In the Aharonov-Bohm effect, a local magnetic field that does not interact directly with the electrons affects the motion of the electrons. This interaction can only take place through space, through the non-local property of the spin.

Electron

In purely formal terms, the riddle can be solved by considering the vector potential A instead of the magnetic field B. The vector potential is a mathematical tool. If you form a specific derivation of the vector potential, the rotation, then you get the magnetic field. With the vector potential as an aid, calculations of the Maxwell equations can be simplified.

Using the vector potential in the Schrödinger equation, one actually gets that the phases of the electron beams through the double slit shift, since the vector potential outside the magnet is not zero.

However, the vector potential is not a physically measurable quantity, but only a mathematical tool. It can therefore not cause any physically measurable effect. How does the physical influence of the magnetic field on the electrons happen?

Since the magnetic field cannot affect the particles directly, it must do so indirectly. The influencing of the particles must therefore take place over space, just as the spins of entangled particles also align themselves accordingly over space during a measurement. In the Aharonov-Bohm effect, the electrons are not entangled, but the spin is part of space and thus not confined to the particle; as experiments on entanglement have shown, the spin of a particle can be influenced even over large distances. The spin is therefore also present in the area of the coil. If the magnetic field in the coil changes, this also affects the spin, even though the particle is located somewhere else. We therefore observe the Aharonov-Bohm effect.

10.2. Quantum Cheshire Cat

In 2013, Yakir Aharonov and his colleagues published another article in which they made a peculiar prediction (Aharonov 2013): With a certain measurement setup, it should be possible to measure a particle in one place, and its state in another. He named this phenomenon "Quantum Cheshire Cat", after the cat from the book "Alice in Wonderland" that almost completely disappears, leaving only its grin behind.

Aharonov and his colleagues chose a clever setup in which a light-ray was split in two and then sent to a beam splitter (BS or polarizing beam splitter PBS) before being measured by different detectors. On the left arm, the beam was not further modified, on the right arm, it went through a phase shifter (PS) and a half wave plate (HWP) which rotates the polarization.

In the Quantum Cheshire Cat effect, you measure a particle in the left arm, but its polarization in the right. The properties of a quantum are not properties of the particle but of space

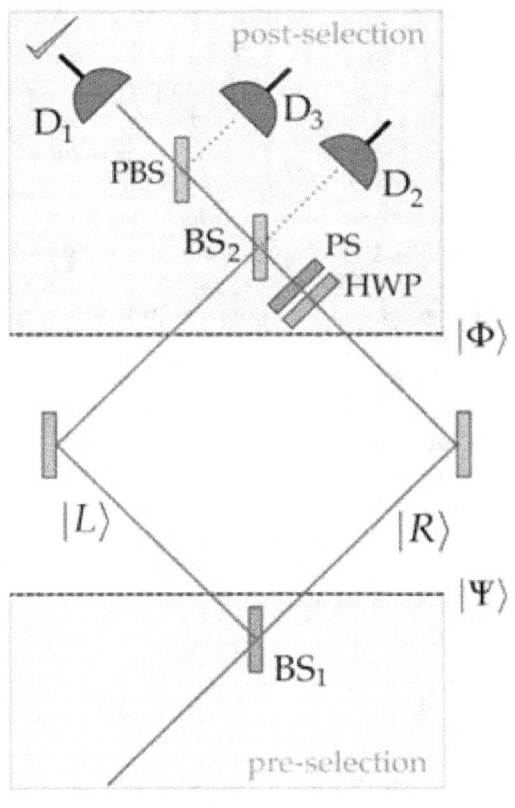

The light wave is prepared in such a way that it can only go through the left branch of the experimental setup. It is then only measured by detector D_1. However, Aharonov and his colleagues predicted that one can still measure a changed polarization, as if the photon also went through the right arm and was rotated there – although this path is excluded.

In this setup, the measurements cannot be performed at the same time, since the measurement of a particle in the left arm interferes with the measurement of polarization (which involves the right arm). So, you can't do these two measurements at the same time, and it seems that it's not possible to verify this prediction.

However, this is possible with a trick, which Aharonov had introduced with colleagues in 1988: The "weak measurement" (Aharonov 1988). Here, not exactly one parameter is measured, whereby the system is changed in the direction of this parameter, but the measurement system is influenced only minimally, and one receives the measurement result only with a certain probability, thus several measurements must be done.

In this case, one measures the polarization by a small rotation of the polarization of the photon and an absorber for the particle, which is almost perfectly transparent. In this way, the system is only minimally affected, and the location of the particle as well as its polarization can be measured at the same time, albeit only with certain probabilities. Such a measurement was performed e.g. by the Kim 2020 and could confirm the Quantum Cheshire Cat effect. The polarization is rotated in the right arm, although the particle passes through the left arm. Property and particle are separated from each other. The polarization is not a property of the particle, but of the space, as could already be assumed after the measurements on entanglement.

10.3. Neutron spin

Another experiment that was published in spring 2022 also fits to these observations. Hartmut Lemmel and his colleagues carried out measurements on the neutron spin (Lemmel 2022). The experimental setup was simple: the neutrons were given a spin in the z-direction using a magnetic field, then the neutron went through a beam splitter. In the lower path was a coil that rotated the spin of the neutron by a small angle as part of a weak measurement, in the upper path was no coil. Then the two paths were brought back together and the neutron's spin was rotated back by a small angle to regain z-alignment.

If the particle had gone through the lower path, then the spin would have had to be reversed by exactly this angle. If the particle had gone through the upper path, then the spin would have had to be rotated back by the angle zero. In fact, one had to rotate the angle by an angle that was between these two values.

The neutron spin is only rotated by an angle on the lower path. Before the measurement, the rotation is reversed, but this requires an angle smaller than the one which rotated the spin on the lower path.

The researchers assumed that this was because the particle had taken both paths at the same time and had therefore been partly influenced by the additional magnetic field and partly not, which is why only a partial rotation was necessary to recover alignment to the z-direction. Therefore, they thought to have proved the

Copenhagen interpretation which assumes that a particle somehow takes both ways. The authors, however, assumed that the spin is a property of the particle.

However, as the measurements on the EPR paradox and Bell's inequality have shown, the spin is a non-local property of the quantum and thus cannot be a property of the local particle. Therefore, another interpretation seems much more convincing: the particle took any of the two paths (but only one), but the magnetic field in the lower path affected the non-local wave of the quantum – and thus also the spin of the particle independent of its path. Thus, a partial rotation of the spin occurred, giving the impression that the particle took both paths at the same time. In this case, however, only the space was changed and thus the spin of the neutrons.

11. On the new picture of the quantum world

The properties of a quantum are not locally related to the measured particle, but propagate non-locally. We can understand them as part of space. The unification between the theory of relativity and quantum mechanics can be pointedly understood in such a way that the theory of relativity tells an object how it should move locally in space, while quantum mechanics describes the non-local properties of space itself connected with the object.

In space, particles cannot propagate faster than the speed of light. Information, signals and energy can therefore only be passed on locally, as required by the theory of relativity.

This speed limit does not apply to space itself. The space itself can expand faster than with the speed of light, as one observes in cosmology, and the properties of the space like spin, polarization and guidance of the particles (i.e. whether it shows particle or wave character), can likewise change non-locally – that's why it can affect other quanta non-locally, what Einstein had called a "spooky action at a distance". Since this action is a property of space and does not take place in space, it does not contradict the laws of the theory of relativity. It does, however, explain what distinguishes the quantum world from the "classical world" that we perceive in everyday life: It is the non-locality of the properties of the quanta. However, since we do not perceive them in everyday life, this is the reason why the properties of the quantum world seem to be so peculiar, and why there are different laws of probability acting in the quantum world than what we know from the classical world, as the measurements on Bell's inequality have shown. The quantum world has no hidden local variables, but it is still deterministic. Both Bohr and Einstein were partly right – and partly wrong.

*

The common denominator between the quantum world and the theory of relativity seems to be space. Put bluntly, relativity describes how objects move in space, while quantum mechanics describes the non-local properties of space itself.

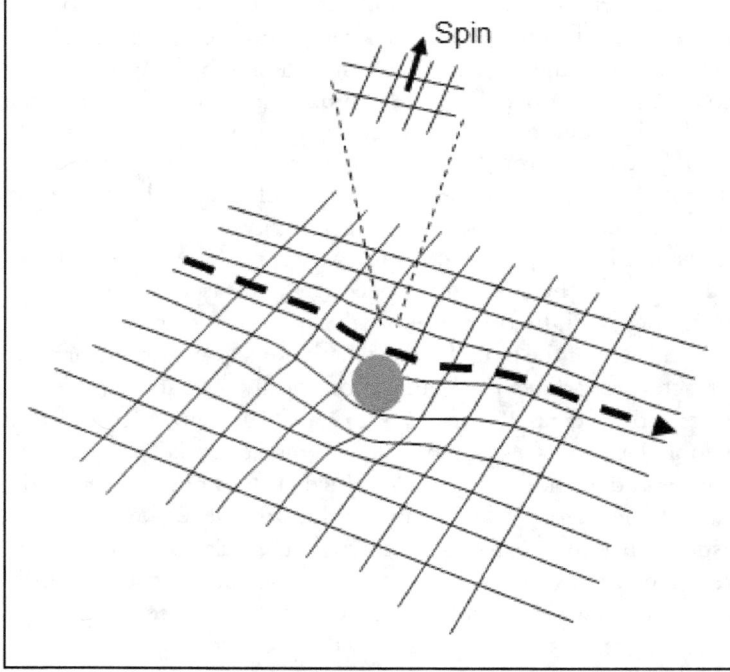

The only theory that provides a distinction between locally moving particles and non-local properties is the de Broglie/Bohm quantum theory. Here a particle is guided through a guiding wave or a quantum potential. De Broglie had shown that this guiding wave must propagate faster than the speed of light, i.e. it is non-local. Bohm's quantum potential is also obviously non-local. This non-locality is the reason why many physicists have rejected this interpretation of quantum mechanics. The fact that the Schrödinger equation also has to be interpreted non-locally, since it is defined in the configuration space and describes the behavior of all particles instantaneously, is often ignored.

However, it is unsatisfactory to follow the approaches of de Broglie and Bohm, since the mathematical description of the guiding wave or the quantum potential is derived from the Schrödinger equation, which had the task of describing the quantum world solely as a wave. The approach thus appears contradictory to the approach of de Broglie and Bohm, both of which assume a local particle in addition to the non-local wave function; a distinction that the Schrödinger equation does not make. In addition, the Schrödinger equation provides imaginary wave functions as a solution, which initially have no physical meaning. Only the square of the absolute value of these wave functions then provides a description for the probability of a state (or the guidance equation derived from it or the quantum potential of the de Broglie-/Bohm-interpretation).

Even if the representation of the quantum world as a local particle and a non-local space resembles very much the de Broglie-/Bohm-representation and one might be tempted to call the non-local space a guiding wave or quantum potential, we want to avoid this for the time being in order to avoid the reference to the questionable Schrödinger equation. The goal of physics must be, based on the description of the quantum world as a local particle and non-local space, to develop formulas that describe the behavior of the particles without first obtaining a non-physical wave function that must be treated further to obtain a physically meaningful statement. The formulas should directly provide a physically meaningful statement and be valid for both electrons and photons.

*

Now, how do we have to imagine a quantum mechanics that takes into account the locality of the particle and the non-locality of space, i.e. the properties? We don't know all the details yet. But we can name some characteristics that it should probably have. The following features are first thoughts on a description of the quantum world, which does not claim to be complete.

*

De Broglie assumed that the particles and the space oscillate with the same frequency. In the case of light, you have photons which have a certain frequency and waves which have the same frequency. The wavelength results then from the speed of light which can be smaller in a medium than in the vacuum. The waves of the light are "pure waves", because the photons do not influence each other. They can be described by a simple wave equation, where in principle the second time derivative corresponds to the second local derivative.

This seems to be different for massive particles, i.e. particles which, unlike the photon, have a rest mass. These particles can no longer be described by a simple wave equation, although their behavior is very much "wave-like". However, it seems that the massive particles are influenced on their way. This is also shown by the Schrödinger equation, where the behavior of one particle depends on the behavior of the other particles – instantaneously; because the wave function is defined in configuration space, which includes the coordinates of all particles. If the parameters of a particle change, then this influences also the other particles. For this, these particles do not even need to be entangled with each other; because a change of the space already takes place if one covers a slit of the double slit.

The big question is how this influence takes place. In the Schrödinger equation one distinguishes between individual particles (which appear in the wave function with their coordinates) and larger objects such as the double slit, which are parts of the potential or the boundary conditions. Since photons do not influence each other, in the case of light we only know the influence of large objects. If we wanted to keep this distinction, then we would have to define the limit for which an object is to be considered large. Otherwise, the task would be to find a wave equation that describes how mass and light particles are influenced by objects in general.

In both cases an even stronger influence seems to take place when the particles are entangled with each other. This entanglement is conveyed through space. The big question is what this looks like – and how the manipulation of the space for entangled particles differs from that for non-entangled particles.

The influence of the particles is instantaneous. A light wave spreads out immediately (and with it the polarization) – the photon then follows "lazily" with the speed of light. The same applies to matter particles. Electrons move through an experimental setup at speeds below the speed of light. If the setup changes, then the influence on the electron changes instantaneously and, as in the delayed choice experiment, you obtain a different result than originally intended.

Space determines non-locally the path a particle can take. The speed at which it travels along the path depends on the speed to which the particle has been accelerated, on the energy that has been given to a particle. This energy causes space-time to warp, which then provides a geodesic along which a particle moves. The property of space modifies this movement. We use the term "space" instead of "space-time". Since the changes in space propagate instantaneously, it does not make sense trying to assign a time to these changes.

Space itself exists without time, because it is non-local. In space things happen instantaneously. Only the movement in the space seems to generate the space-time – and with it the time which is connected with the space in such a strange way as Einstein has explained in his theory of relativity.

The path of a particle can be "wavelike" in space. It need not be a straight line. Some critics think that the way of a particle cannot change so simply, that this would violate the law of conservation of momentum, after all a particle changes its direction and thus also its momentum. But already with entropy one had to accept that the rule that entropy increases in a closed system does not hold absolutely, but only statistically. That the momentum cannot be a constant in the quantum world is also shown by the phenomenon of diffraction, where light – and thus the photon – goes "around the corner". It changes its direction – and thus its momentum. Only in the macroscopic world, where one averages over billions of particles, the momentum seems to be a constant. But even in the macroscopic world this is not really true, as objects travel along a geodesic, which do not have to be a linear line, but can be warped and bent and thus would not assure a conservation of momentum as we usually understand it.

*

The properties of space do not only include that it modifies a particle's trajectory by guiding it, but space also possesses properties such as polarization and spin that are normally ascribed to the particle. However, the EPR paradox and the investigations regarding Bell's inequality can only be explained by assuming that these are also properties of space. This is the only way a non-local spread is possible.

The question is which other properties are properties of space. If the magnetic moment of the spin is part of space and not of the particle, then one should also assume that the electric charges are also part of space; after all, electricity and magnetism are just two sides of the same coin, which are summarized in electromagnetism.

That the electric charge is part of the non-local space and not part of the particle also follows from its behavior at higher velocities: it does not change. Unlike a mass, which according to the theory of relativity increases as the particle's speed increases because it is moving in space, the charge remains unchanged; this indicates that it is not part of the particle moving in space (otherwise it should also be affected by the movement in space), but a property of non-local space.

Mass increases the faster a particle moves, but the charge remains unchanged. This could be because the charge is a property of space and therefore does not change as the particle moves in space.

Mass

resting   moving

Charge

However, with the help of electric charges, energy can be stored (think of a capacitor). If the charges are part of the space – is the energy also stored in the space (i.e. as a property of space)?

The energy is certainly stored in the space between the charges. But since the rules of relativity apply to energy and mass, electrical energy cannot be a property of space; it must reside in space, in space-time, just as particles do. After all, energy and mass are equivalent according to Einstein.

Since the electric energy is finally the cause for the fact that particles exert an electric force on each other, it follows that the force is also a local effect which takes place in the space, just as masses form space-time locally and photons and electrons transfer energy and mass locally.

*

The understanding of the quantum world as a world in which the behavior is described by the properties of space and thus also by the other particles can also help to understand the so-called "virtual particles". In order to be able to describe exactly the interactions in the quantum world, one has to assume the existence of virtual particles that influence the "pure behavior" of two particles with each other in order to obtain a mathematically correct description. Without the influence of virtual particles, which arise spontaneously and exist only for a very short time but still influence the "real" particles, quantum theory cannot correctly describe the interactions of particles.

The existence of virtual particles seems to be curious anyway: How can a particle be created spontaneously, and thus energy, if the theorem of conservation of energy categorically excludes the spontaneous creation of energy? The solution is found in the Heisenberg uncertainty principle: If these virtual particles exist only briefly enough, then they fall under the measurement uncertainty and if one cannot measure something, so the Copenhagen interpretation tells us, then it can be anything. It can be a wave or a particle or something even more bizarre, it can also exist and influence other particles – just not long enough for us to measure it.

Why the quantum theory according to the Copenhagen interpretation, which only wants to accept measurable things, nevertheless assumes the existence of virtual particles at all, which by definition are not measurable, seems contradictory, but quantum theory does not manage to describe the interactions of particles mathematically in any other way. The interactions show deviations from the behavior one would expect if only the particles that actually exist interacted.

However, these deviations can also be understood if one considers that all particles are connected to one another across space. Changes in other particles also affect the interacting particles, just as water molecules change the trajectory of a Brownian particle. This inevitably leads to deviations from the "pure" interaction – only one no longer has to assume that virtual particles exist, the existence of which cannot be proven anyway.

But isn't there an indirect proof of virtual particles, using the Casimir effect? The Casimir effect was named after Hendrik Casimir, who predicted it in 1948. The effect is the reason that a force acts on two parallel, conductive plates in the vacuum, and this force presses the plates together (Casimir 1948); because both outside of the plates and between them virtual particles are being formed. However, between the plates only particles whose wavelengths are an even fraction of the distance can form, while outside the plates all possible particles can form. Thus, there are more virtual particles outside than between the plates, the particles outside therefore exert a greater pressure, and the plates are pressed together. This effect has also been confirmed experimentally, e.g. in 1997 with an accuracy of five percent relative to the prediction (Lamoreaux 1997).

However, the matter is not so clear-cut. This effect can also be explained by van der Waals forces, i.e. by relativistic forces between charges and currents (this is how Casimir had originally explained the effect, see Jaffe 2005), so that the effect is not a clear proof for the existence of virtual particles. Their existence is still hypothetical – and maybe they do not exist at all.

*

Not only does space seem to affect a particle, but a particle must also affect space. This becomes clear when considering particles such as electrons. De Broglie showed that the wavelength associated with a particle is inversely proportional to its momentum. The greater the momentum, and thus the greater the kinetic energy of a particle, the smaller the wavelength. If you change the energy of a particle, this also has an influence on space and the guidance properties of space, which guide a particle as if it were guided by a wave. De Broglie had also deduced that the oscillations of the particle and of space (the guiding wave) are in phase with one another. The particle imposes a form on space via its kinetic or vibrational energy, and with this form space guides the particle. Space and particles influence each other.

*

The particle is not only influenced by space. The particle also influences space. If the kinetic energy and thus the momentum of the particle changes, then the wavelength of the particle changes, i.e. the way in which space guides the particle. According to de Broglie we have:

$$p = \frac{h}{\lambda}$$

The momentum p is inversely proportional to the wavelength λ. In addition, de Broglie proposed that the energy of a particle corresponds to the vibration of the particle - which is in phase with the vibration of space. That's how a particle could influence space through its energy – which then in turn directs the particle.

In order to understand the quantum world and to combine it with the theory of relativity, it no longer seems necessary to quantize space-time, but the behavior of the quantum world results from understanding space as the carrier of the properties of the quanta. The theory of relativity describes the behavior of the space-time, if masses are found in it (respectively energy), and how these masses are led through the space-time changed by the masses.

The quantum theory should describe the properties of the space, which then influence the behavior of the particles – whereby the particles can also influence the space again.

The connecting factor between quantum world and general relativity is not a quantized space-time, but the non-local space. Only a quantum theory, which describes the local behavior of the particles and the non-local behavior of their properties, will be able to serve as a basis to unite the quantum theory with the theory of relativity – and thus to obtain a "theory of everything".

Mathematical Appendix

Appendix A: De Broglie's Discoveries

On the one hand, de Broglie calculated the frequency based on Einstein's famous formula and for the quantum mechanical relationship between energy and frequency:

$$E = h\nu = mc^2$$

In a system moving with velocity v, the mass increases according to:

$$E' = m'c^2 = \frac{mc^2}{\sqrt{1 - \left(\frac{v}{c}\right)^2}}$$

From this it follows that the frequency also increases with the speed:

$$\nu' = \frac{E'}{h} = \frac{mc^2}{h\sqrt{1 - \left(\frac{v}{c}\right)^2}} = \frac{\nu_0}{\sqrt{1 - \left(\frac{v}{c}\right)^2}}$$

Likewise, time increases at high speed (it slows down). Since frequency is the reciprocal of time, it also holds that frequency decreases as speed increases:

$$\nu_1' = \nu_0 \sqrt{1 - \left(\frac{v}{c}\right)^2}$$

Thus, you get a different behavior for the frequency. De Broglie concluded that these formulas must have different physical meanings.

The formula

$$\nu_1' = \nu_0 \sqrt{1 - \left(\frac{v}{c}\right)^2}$$

Describes according to de Broglie the motion of a particle with velocity v, which performs an internal oscillation with this frequency. This velocity is equal to the group velocity of a particle, because with the particle also information is transmitted. This velocity is also always smaller than the speed of light, so that there is no contradiction to Einstein's theory of relativity.

The formula

$$\nu' = \frac{\nu_0}{\sqrt{1 - \left(\frac{v}{c}\right)^2}}$$

on the other hand, which he obtained from the quantum mechanical formula for energy, describes the frequency of a phase wave according to de Broglie. For we have:

$$\nu_1' = \nu'\left(1 - \frac{v^2}{c^2}\right)$$

i.e. a wave with the frequency ν' and the phase velocity

$$v_{Ph} = \frac{c^2}{v}$$

which remains in phase with the internal oscillation of the particle. This phase velocity is always significantly larger than the speed of light! However, with the phase alone no energy is connected and also no information is transferred. Therefore, this is not in contradiction to the theory of relativity. This phase belongs to a wave which propagates in the space and leads the slower moving particle.

De Broglie showed that this phase wave can actually describe the movement of a particle by showing an equivalence of Fermat's principle for the propagation of waves with Maupertuis' principle for the propagation of particles.

Mathematically, Fermat's principle can be formulated in such a way that the sum over all phases φ of a wave between two points P and Q must be constant:

$$\int_P^Q d\varphi = const.$$

Or, to point out the extremal principle, that small variations must be zero:

$$\delta \int_P^Q d\varphi = 0$$

This is the general formulation of Fermat's principle in optics. Incidentally, the constant is called the action (S). It has the unit energy times time.

In classical mechanics, a very similar extremal principle applies, the Maupertuis principle, which was formulated in the 18th century by Pierre de Maupertuis. While Fermat had looked at the shortest path of waves, Maupertuis considered the shortest path taken by particles. Similarly to waves, the action must become minimal for particles. For particles, the equation is:

$$S = \int_P^Q mv \, ds$$

Here m is the mass, v is the velocity of the particle, and ds represents the path segments over which the sum is taken. For small variations, this action must also be zero.

The small phase changes $d\varphi$ between waves that have frequency v on average are caused by the fact that they are on a slightly different path ds, which they travel with phase velocity v_{Ph}. Thus, we have for the phase change:

$$d\varphi = 2\pi v \frac{ds}{v_{Ph}}$$

For the phase velocity of the particle, we have:

$$v_{Ph} = \frac{c^2}{v}$$

Where v is the group velocity. Thus, we have

$$d\varphi = 2\pi v \frac{v ds}{c^2}$$

According to Planck the energy is proportional to the frequency, at the same time Einstein's famous formula $E=mc^2$ is valid for the energy. So we have:

$$v = \frac{E}{h} = \frac{mc^2}{h}$$

Substituting this into the formula for the change in phase, we have

$$d\varphi = \frac{2\pi}{h} mv \, ds$$

We can therefore write the action according to Fermat as

$$S = \int_P^Q d\varphi = \frac{2\pi}{h} \int_P^Q mv \, ds$$

Which is similar to Maupertuis' extremal principle, except for a constant that is of no interest when forming extremals:

$$S = \int_P^Q mv \, ds$$

Matter particles also show wave and particle properties like photons, however these properties might be caused.

Appendix B: Derivation of the basic equations of the de Broglie-/Bohm-mechanics

The starting point for Bohm's considerations was Schrödinger's wave equation, as it was known that it can describe the observations well. He solved the Schrödinger equation

$$i\hbar \frac{d\Psi}{dt} = -\left(\frac{\hbar^2}{2m}\right)\nabla^2\Psi + V(r)\Psi$$

with the approach:

$$\Psi = Re^{iS/\hbar}$$

Here R is the amplitude of the wave, S is its phase. If we insert this approach into the Schrödinger equation and write down the imaginary and real parts separately, then we get:

$$\frac{\partial R}{\partial t} = -\frac{1}{2m}[R\nabla^2 S + 2\nabla R \, \nabla S]$$

$$\frac{\partial S}{\partial t} = -\left[\frac{(\nabla S)^2}{2m} + V(r) - \frac{\hbar^2}{2m}\frac{\nabla^2 R}{R}\right]$$

The square of the absolute value of the wave function is $|\Psi|^2 = R^2$, and this is equal to the probability density ϱ. If we derive this with respect to time, then we have

$$\frac{\partial \rho}{\partial t} = 2R\frac{\partial R}{\partial t}$$

This gives for the first equation:

$$\frac{\partial \rho}{\partial t} + \nabla\left[\rho\frac{\nabla S}{m}\right] = 0$$

This resembles a continuity equation for ϱ. A continuity equation has the general form:

$$\frac{\partial \rho}{\partial t} + \nabla j = 0$$

Here ϱ is the density of a quantity, such as mass or charge, and j is the current density. The change in density over time therefore corresponds to the change in current density in space. A mass that flows into a volume in a given period of time is equal to the mass that leaves the volume. If the continuity equation holds for mass, then we have a conservation of mass. The same applies to charges, the number of particles or energy.

The current density is equal to the density multiplied by the velocity:

$$j = \rho v$$

Thus, according to Bohm's continuity equation, the velocity of a quantum is:

$$v = \frac{\nabla S}{m}$$

This had already been derived by de Broglie in 1927.
This interpretation is also obtained from the second equation. One sees this as follows: In general, the variation of R is smaller than that of S, so the term

$$-\frac{\hbar^2}{2m}\frac{\nabla^2 R}{R}$$

in the second equation can be neglected in a first approximation. It follows:

$$\frac{\partial S}{\partial t} + \frac{(\nabla S)^2}{2m} + V(r) = 0$$

The British physicist William Rowan Hamilton and the German mathematician Carl Gustav Jacobi had developed an equation for classical mechanics, which had the form

$$\frac{\partial S}{\partial t} + \frac{p^2}{2m} + V(r) = 0$$

Thus, one can again identify the momentum p with

$$p = \nabla S$$

and the velocity with

$$v = \frac{\nabla S}{m}$$

According to Bohm, the term that we originally neglected as small is the quantum potential

$$U_{quant} = -\frac{\hbar^2}{2m}\frac{\nabla^2 R}{R}$$

which ultimately determines the movement of the particles. Bohm saw the quantum potential as the actual physical quantity that determines the trajectory of a quantum particle.

De Broglie, on the other hand, originally set himself the goal of developing a new dynamic for quantum mechanics. He wanted to give an equation of motion. For him, therefore, the basic formula was the formula

$$\frac{dQ}{dt} = v = \frac{\nabla S}{m}$$

which was derived with the special ansatz

$$\Psi = R e^{iS/\hbar}$$

In general, it has the form:

$$\frac{dQ}{dt} = \frac{\hbar}{m} Im\left(\frac{\nabla\Psi}{\Psi}\right)$$

The "Im" means that you only look at the imaginary part of the fraction.

This formula describes the path or trajectory of the particle.

Today, in Bohmian mechanics, one rather follows de Broglie's approach. One calculates from the Schrödinger equation

$$i\hbar\frac{d\Psi}{dt} = -\left(\frac{\hbar^2}{2m}\right)\nabla^2\Psi + V(r)\Psi$$

the wave function Ψ, and then using the wave function, the particle trajectory according to the formula:

$$\frac{dQ}{dt} = \frac{\hbar}{m} Im\left(\frac{\nabla\Psi}{\Psi}\right)$$

Thus, the wave function "guides" the particle through the world.

Appendix C: Derivation of Bell's inequality

In deriving the inequality, Bell envisioned a modified EPR experiment as suggested by David Bohm. Instead of momentum and position, Bohm considered two electrons each having a spin of ½. Together they have spin 0, so one particle has spin ½ and the other particle spin -½, but you don't know which particle has which spin. You only find this out when you measure the spin of one particle – which determines the spin of the other particle.

To simplify the calculation, we now assume that the particles have either spin +1 or -1.

We denote the measurement of the first particle with A, the measurement of the second particle with B. If the particles are entangled, then the measurement of A determines the measurement result of B (if A is +1, for example, then we have B=-1). Einstein now claims that the measurement result not only depends on the wave function, but also on an additional parameter that determines the measurement results in advance. Bell denoted this parameter with the Greek letter λ and left completely open whether this is a number or a function. This is also completely irrelevant for his proof.

So now you have a measurement A that measures the spin in a direction a and depends on a hidden parameter λ, the same applies to measurement B:

$$A(a,\lambda) = \pm 1, \quad B(b,\lambda) = \pm 1$$

Quantum mechanically, we only experience probabilities. The probability for the measurement in the directions a, b is calculated as the product of the probabilities for the individual measurements

A and B multiplied by the distribution of the hidden parameters $\varrho(\lambda)$ (the density function):

$$P(a,b) = \int d\lambda \rho(\lambda) A(a,\lambda) B(b,\lambda)$$

Bell now looked at another direction in which the measurement can be carried out. Looking at the same direction and opposite spin, we have perfect anticorrelation: $A(a,\lambda) = - B(a,\lambda)$. It follows that:

$$P(a,b) = - \int d\lambda \rho(\lambda) A(a,\lambda) A(b,\lambda)$$

For a third measurement in direction c, the following generally applies:

$$P(a,b) - P(a,c)$$
$$= - \int d\lambda \rho(\lambda) [A(a,\lambda) A(b,\lambda)$$
$$- A(a,\lambda) A(c,\lambda)]$$

The square of the absolute value of the probability is one, i.e. $|A(b,\lambda)|^2 = 1$. If you expand the second summand with this factor, then you get

$$P(a,b) - P(a,c)$$
$$= - \int d\lambda \rho(\lambda) A(a,\lambda) A(b,\lambda) [1$$
$$- A(b,\lambda) A(c,\lambda)]$$

Now $A(a,\lambda) = \pm 1$ and $A(b,\lambda) = -B(b,\lambda)$. Taking the absolute value of the equation and considering that the probabilities are less than or equal to one, we get the following inequality:

$$|P(a,b) - P(a,c)| \leq \int d\lambda \rho(\lambda) [1 - A(b,\lambda) A(c,\lambda)]$$

The density function is normalized to one, and one obtains for the integral:

$$|P(a,b) - P(a,c)| \leq 1 - P(b,c)$$

This is Bell's inequality. It is sometimes written without the absolute value bars in the form

$$P(a,b) - P(a,c) + P(b,c) \leq 1$$

Any physical equation with local parameters must satisfy this equation for the probabilities of measurements. Laws of classical physics do this.

In quantum mechanics, the probability depends on the cosine between the two measurement directions a and b.

$$P(a,b) = \cos 2\alpha_{ab}$$

Let us now assume that two measuring directions a and c have an angle of 60°, and the measuring direction b lies on the angle bisector between a and c (i.e. here is an angle of 30°). Then we have:

$$P(a,b) - P(a,c) + P(b,c)$$

$$= \cos 2\alpha_{ab} - \cos 2\alpha_{ac} + \cos 2\alpha_{bc}$$

$$= \frac{1}{2} - \left(-\frac{1}{2}\right) + \frac{1}{2} = \frac{3}{2}$$

and this does not satisfy Bell's inequality, as it is larger than 1.

Appendix D: Derivation of the field equations of general relativity

According to the general theory of relativity, space-time has a Riemannian geometry, i.e. at every point in space there is a tangent space that locally resembles a Minkowski space. Therefore, the mathematical nomenclature for the Minkowski space will be derived first and then it will be generalized for generally curved spaces.

According to Minkowski, the invariant of space-time is the space-time interval:

$$s^2 = -(c\,\Delta t)^2 + (\Delta x)^2 + (\Delta y)^2 + (\Delta z)^2$$

A slightly more compact notation is to denote the coordinates with the letter x, and give them a superscript Greek letter (μ) running from 0 to 3, where:

$$x^0 = ct$$
$$x^1 = x$$
$$x^2 = y$$
$$x^3 = z$$

If one uses a Latin letter instead of a Greek letter, then the convention is that it only runs from 1 to 3 and thus only lists the spatial dimensions.

As a rule, for the sake of simplicity, units are used where we define

$$c = 1$$

One can then define the Minkowski metric as follows:

$$\eta_{\mu\nu} = \begin{pmatrix} -1 & 0 & 0 & 0 \\ 0 & 1 & 0 & 0 \\ 0 & 0 & 1 & 0 \\ 0 & 0 & 0 & 1 \end{pmatrix}$$

This allows to simply write the space-time interval, the line element, as:

$$s^2 = \eta_{\mu\nu} \Delta x^\mu \Delta x^\nu$$

Einstein's sum convention applies here: The summation is done over upper and lower indices. Since the Minkowski metric has non-zero values only in the diagonals, only the square numbers remain.

Minkowski's space-time is Euclidean in its coordinates. If one imagines the general space-time as curved, then there can be a tangent space at this space, which lies tangential to the curved space-time at a point (in three dimensions one can imagine this as a surface on a curved surface).

With the basis vectors of plane space-time

$$\hat{e}_{(\mu)}$$

a vector A can be written as:

$$A = A^\mu \hat{e}_{(\mu)}$$

For this vector space one can define a dual vector space that contains the basis vectors

$$\hat{\theta}^{(\nu)}$$

The basis vectors of the dual vector space are defined in such a way that:

$$\hat{\theta}^{(\nu)}\left(\hat{e}_{(\mu)}\right) = \delta^\nu_\mu$$

Vectors of the dual vector space are thus:

$$\omega = \omega_\mu \hat{\theta}^{(\mu)}$$

The vectors of the original vector space are also called contravariant vectors (upper indices), those of the dual vector space are called covariant vectors (lower indices).

The product of a covariant vector and a "normal" vector is simply a scalar. You can also think of the "normal" vector as a column vector and the covariant vector as a row vector:

$$V = \begin{pmatrix} V^1 \\ V^2 \\ \cdot \\ \cdot \\ \cdot \\ V^n \end{pmatrix} ; \; \omega = (\omega_1 \omega_2 \ldots \omega_n)$$

The product is therefore:

$$\omega(V) = (\omega_1 \omega_2 \ldots \omega_n) \begin{pmatrix} V^1 \\ V^2 \\ \cdot \\ \cdot \\ \cdot \\ V^n \end{pmatrix} = \omega_i V^i$$

A generalization of vectors and covariant vectors is the tensor. It is practically nothing more than a mapping consisting of a collection of vectors and covariant vectors. A tensor of type (or rank) (k, l) consists of k "normal" vectors (upper indices) and l covariant vectors (lower indices).

The basis for a (k, l)-tensor consists of the tensor product of the individual basis vectors:

$$\hat{e}_{(\mu_1)} \otimes \cdots \otimes \hat{e}_{(\mu_k)} \otimes \hat{\theta}^{(\nu_1)} \otimes \cdots \otimes \hat{\theta}^{(\nu_l)}$$

A tensor can thus be written as:

$$T = T^{\mu_1 \cdots \mu_k}{}_{\nu_1 \cdots \nu_l} \, \hat{e}_{(\mu_1)} \otimes \cdots \otimes \hat{e}_{(\mu_k)} \otimes \hat{\theta}^{(\nu_1)} \otimes \cdots \otimes \hat{\theta}^{(\nu_l)}$$

Alternatively, the components can also be defined by letting the tensor act on the basis vectors and the dual basis vectors:

$$T^{\mu_1 \cdots \mu_k}{}_{\nu_1 \cdots \nu_l} = T(\hat{\theta}^{(\mu_1)}, \dots, \hat{\theta}^{(\mu_1)}, \hat{e}_{(\nu_1)}, \dots, \hat{e}_{(\nu_l)})$$

A simple tensor is a $(1, 1)$-tensor that is simply a matrix. Its effect on a vector and a covariant vector is given by the well-known matrix multiplication:

$$M(\omega, V) = (\omega_1 \omega_2 \dots \omega_n) \begin{pmatrix} M^1{}_1 & M^1{}_2 & \dots & M^1{}_n \\ M^2{}_1 & M^2{}_2 & \dots & M^2{}_n \\ \cdot & \cdot & \dots & \cdot \\ \cdot & \cdot & \dots & \cdot \\ \cdot & \cdot & \dots & \cdot \\ M^n{}_1 & M^n{}_2 & \dots & M^n{}_n \end{pmatrix} \begin{pmatrix} V^1 \\ V^2 \\ \cdot \\ \cdot \\ \cdot \\ V^n \end{pmatrix}$$

$$= \omega_i M^i{}_l V^j$$

The inverted metric (upper indices) is defined by:

$$\eta^{\mu\nu}\eta_{\nu\rho} = \eta_{\rho\nu}\eta^{\nu\mu} = \delta^\nu_\mu$$

We also have the Levi-Civita-Tensor:

$$\epsilon_{\mu\nu\rho\sigma} \begin{cases} +1, \textit{if } \mu\nu\rho\sigma \textit{ is an even permutation of } 0123 \\ -1, \textit{if } \mu\nu\rho\sigma \textit{ is an odd permutation of } 0123 \\ \quad\quad 0, \textit{else} \end{cases}$$

With tensors, we have the operation of contraction, in which one turns a (k, l)-tensor into a $(k-1, l-1)$ tensor. This is done by summing over an upper and a lower index. Of course, this depends on the order of the indices.

$$S^{\mu\rho}{}_\sigma = T^{\mu\nu\rho}{}_{\sigma\nu}$$

The metric and the inverse metric can be used to raise or lower indexes on a tensor:

$$T^{\alpha\beta\mu}{}_\delta = \eta^{\mu\gamma} T^{\alpha\beta}{}_{\gamma\delta}$$

$$T_\mu{}^\beta{}_{\gamma\delta} = \eta_{\mu\alpha} T^{\alpha\beta}{}_{\gamma\delta}$$

In Euclidean geometry, the partial derivative of a (k, l) tensor is a $(k, l+1)$ tensor

$$T_\alpha{}^\mu{}_\nu = \partial_\alpha R^\mu{}_\nu$$

which satisfies the Lorentz transformations. However, this does not apply in a general space, there we will have to define a "covariant derivative" that replaces the partial derivative.

This brings us to the general spaces. Here the tangent space is defined as the space with directed derivatives along a curve through a point p. The basis vectors are thus generally defined as partial derivation operators:

$$\hat{e}_{(\mu)} = \partial_\mu = \frac{\partial}{\partial x^\mu}$$

The basis vectors in a new coordinate system can be easily obtained using the chain rule:

$$\partial_{\mu'} = \frac{\partial x^\mu}{\partial x^{\mu'}} \partial_\mu$$

Likewise, one obtains the new vector, which should be unchanged by the new basis:

$$V^\mu \partial_\mu = V^{\mu'} \partial_{\mu'} = V^{\mu'} \frac{\partial x^\mu}{\partial x^{\mu'}} \partial_\mu$$

Since the derivation matrices are simply inverses of each other, we have:

$$V^{\mu'} = \frac{\partial x^{\mu'}}{\partial x^\mu} V^\mu$$

In dual, cotangent space, the gradients of the coordinate functions form the basis, because we have

$$dx^\mu(\partial_\nu) = \frac{\partial x^\mu}{\partial x^\nu} = \delta^\mu_\nu$$

The following then applies analogously to the transformation rules:

$$dx^{\mu'} = \frac{\partial x^{\mu'}}{\partial x^\mu} dx^\mu$$

or for the components of the dual cotangent vector:

$$\omega_{\mu'} = \frac{\partial x^\mu}{\partial x^{\mu'}} \omega_\mu$$

For the metric in general space, a new name is introduced:

$$g_{\mu\nu}$$

The inverse metric is then defined via:

$$g^{\mu\nu} g_{\mu\sigma} = \delta^\mu_\sigma$$

The metric performs several tasks:
- It allows a distinction between past and future
- It allows the calculation of path lengths and proper time

- It replaces Newton's gravitational field
- It determines causality because it defines the speed of light as the highest speed for signals

The line element (or the metric, since dx^μ is only a dual basis vector) is then:

$$ds^2 = g_{\mu\nu}dx^\mu dx^\nu$$

In a general metric, the Levi-Civita symbol is defined by:

$$\tilde{\epsilon}_{\mu\nu\rho\sigma} \begin{cases} +1, \text{if } \mu\nu\rho\sigma \text{ is an even permutation of } 0123 \\ -1, \text{if } \mu\nu\rho\sigma \text{ is an odd permutation of } 0123 \\ 0, \text{else} \end{cases}$$

This is not a tensor as it changes under coordinate transformations. The Levi-Civita tensor is defined as:

$$\epsilon_{\mu_1\mu_2...\mu_n} = \sqrt{|g|}\tilde{\epsilon}_{\mu_1\mu_2...\mu_n}$$

If we want to calculate the derivative, we have to consider that the direction in which a vector is transported in parallel depends on the curvature. Let's take two arrows that are perpendicular to the equator of the earth. One arrow is on the zero meridian, the other at 90°. If both are now transported in parallel along the meridian, they meet at the North Pole at a right angle. If you were to move an arrow further around the triangle, for example the one at 0°, you would get a vector at the starting point that is inclined to the original vector. This parallel transport along a closed curve changes the direction of the vector, with the deviation of its direction being a measure of the curvature of the surface. Therefore, one cannot simply take the partial derivative as the direction on an arbitrarily shaped surface, but one must define a covariant derivative. To assure that this does not become too complicated, one defines the covariant derivative as a partial derivative plus a correction:

$$\nabla_\mu V^\nu = \partial_\mu V^\nu + \Gamma^\nu_{\mu\lambda} V^\lambda$$

For a covariant vector we have:

$$\nabla_\mu \omega_\nu = \partial_\mu \omega_\nu - \Gamma^\lambda_{\mu\nu} \omega_\lambda$$

We define that the connection should be torsion-free, i.e.

$$\Gamma^\lambda_{\mu\nu} = \Gamma^\lambda_{[\mu\nu]}, with \; \Gamma^\lambda_{\mu\nu} - \Gamma^\lambda_{\nu\mu} = 2\Gamma^\lambda_{[\mu\nu]}$$

In addition, a connection should be metrically compatible if the covariant derivative of the metric with respect to that connection is zero everywhere:

$$\nabla_\rho g_{\mu\nu} = 0$$

If we write this using to the definition of the covariant derivative, then we have:

$$\nabla_\rho g_{\mu\nu} = \partial_\rho g_{\mu\nu} - \Gamma^\lambda_{\rho\mu} g_{\lambda\nu} - \Gamma^\lambda_{\rho\nu} g_{\mu\lambda} = 0$$

We take this equation and permute the three indices. This gives us three equations. The permuted ones are subtracted from the first, consider the symmetries and finally we obtain for the so-called Christoffel symbol:

$$\Gamma^\sigma_{\mu\nu} = \frac{1}{2} g^{\sigma\rho} \left(\partial_\mu g_{\nu\rho} + \partial_\nu g_{\rho\mu} - \partial_\rho g_{\mu\nu} \right)$$

One also sometimes writes:

$$\Gamma^\sigma_{\mu\nu} = \left\{ {\sigma \atop \mu\nu} \right\}$$

Particles move along a geodesic. If the path is given by x(λ), then the geodesic equation is:

$$\frac{d^2 x^\mu}{d\lambda^2} + \Gamma^\mu_{\rho\sigma} \frac{dx^\rho}{d\lambda} \frac{dx^\sigma}{d\lambda} = 0$$

If a particle is moving slowly and the field is static and weak, we get Newton's equation of motion. The parameter of the geodesic may be the proper time τ. Slow movement means:

$$\frac{dx^i}{d\tau} \ll \frac{dt}{d\tau}$$

This simplifies the geodetic equation to:

$$\frac{d^2 x^\mu}{d\tau^2} + \Gamma^\mu_{00} \left(\frac{dt}{d\tau}\right)^2 = 0$$

Since the field is static, the Christoffel symbol is simply:

$$\Gamma^\mu_{00} = \frac{1}{2} g^{\mu\lambda} (\partial_0 g_{\lambda 0} + \partial_0 g_{0\lambda} - \partial_\lambda g_{00})$$

$$= -\frac{1}{2} g^{\mu\lambda} \partial_\lambda g_{00}$$

In a weak gravitational field, we can decompose the metric into a Minkowski part plus a weak perturbation:

$$g_{\mu\nu} = \eta_{\mu\nu} + h_{\mu\nu}$$

With this we get:

$$\Gamma^\mu_{00} = -\frac{1}{2} \eta^{\mu\lambda} \partial_\lambda h_{00}$$

The geodesic equation is thus:

$$\frac{d^2x^\mu}{d\tau^2} = \frac{1}{2}\eta^{\mu\lambda}\partial_\lambda h_{00}\left(\frac{dt}{d\tau}\right)^2$$

If we consider only the time component on the left-hand side, the derivation is zero. Only the space components remain, which in the Minokwski case, however, represent a 3x3 identity matrix. So we have:

$$\frac{d^2x^i}{d\tau^2} = \frac{1}{2}\left(\frac{dt}{d\tau}\right)^2\partial_i h_{00}$$

If we divide both sides by the derivation of time with respect to the proper time (squared), then we practically replace the derivation with respect to the proper time by the derivation with respect to time on the left-hand side and obtain:

$$\frac{d^2x^i}{dt^2} = \frac{1}{2}\partial_i h_{00}$$

In Newtonian mechanics, acceleration is proportional to a force or the negative derivation of a potential:

$$a = -\nabla\Phi$$

The approximation of the geodesic equation thus corresponds to Newton's equation of motion if we identify (we will need this expression again later):

$$h_{00} = -2\Phi$$

Let's now move to the derivation of the field equation.
When a vector moves in a closed path on a curved surface, it changes its direction. The change in direction is described by a curvature tensor known as the Riemann tensor. The following formula has been derived for this:

$$R^{\rho}{}_{\sigma\mu\nu} = \partial_{\mu}\Gamma^{\rho}_{\nu\sigma} - \partial_{\nu}\Gamma^{\rho}_{\mu\sigma} + \Gamma^{\rho}_{\mu\lambda}\Gamma^{\lambda}_{\nu\sigma} - \Gamma^{\rho}_{\nu\lambda}\Gamma^{\lambda}_{\mu\sigma}$$

The Riemann tensor is antisymmetric in its last two indices:

$$R^{\rho}{}_{\sigma\mu\nu} = -R^{\rho}{}_{\sigma\nu\mu}$$

For the derivation of the Riemann tensor, the Bianchi identity is known:

$$\nabla_{\lambda}R_{\rho\sigma\mu\nu} + \nabla_{\rho}R_{\sigma\lambda\mu\nu} + \nabla_{\sigma}R_{\lambda\rho\mu\nu} = \nabla_{[\lambda}R_{\rho\sigma]\mu\nu} = 0$$

The Riemann tensor can be contracted to the Ricci tensor:

$$R_{\mu\nu} = R^{\lambda}{}_{\mu\lambda\nu}$$

One can imagine the effect of the Ricci tensor as a change in volume that a sphere experiences when it is transported along a geodesic. Let's imagine a cloud of dust around the earth. First, the earth is not there, space-time is flat. Then the earth appears, space-time is curved, the dust cloud around the earth is attracted and thus changes its volume. This change in volume is described by the Ricci tensor.

The Ricci tensor is symmetric:

$$R_{\mu\nu} = R_{\nu\mu}$$

With the help of the metric one can further contract the Ricci tensor to the Ricci scalar:

$$R = R^{\mu}{}_{\mu} = g^{\mu\nu}R_{\mu\nu}$$

Contracting the Bianchi identity twice gives:

$$0 = \nabla^{\mu}R_{\rho\mu} - \nabla_{\rho}R + \nabla^{\nu}R_{\rho\nu}$$

Or

$$\nabla^\mu R_{\rho\mu} = \frac{1}{2}\nabla_\rho R$$

Defining the Einstein tensor as

$$G_{\mu\nu} = R_{\mu\nu} - \frac{1}{2}Rg_{\mu\nu}$$

then we see that the twice contracted Bianchi identity is equivalent to:

$$\nabla^\mu G_{\mu\nu} = 0$$

The Einstein tensor can be understood as a measure of the excess radius. In Euclidean space, the surface area of a sphere is $4\pi r^2$. In curved space, the surface area is smaller than one would expect based on the radius, so there is an "excess radius". The Einstein tensor is a measure of this excess radius.

Now we have the tools to derive the Einstein equation for general relativity. The idea is that energy and mass deform space. The energy distribution is described by the energy-momentum tensor:

$$T_{\mu\nu}$$

The Riemann tensor describes the curvature of space. However, the Riemann tensor has four indices, the energy-momentum tensor only two. However, with the Ricci tensor there is a version of the Riemann tensor with two indices. So, one could assume that the Ricci tensor is proportional to the energy-momentum tensor, with the proportionality constant $\varkappa$:

$$R_{\mu\nu} = \kappa T_{\mu\nu}$$

However, Einstein assumes that curved space is not different from flat space on a small scale. In this case, the conservation of energy applies. This means, the conservation of energy should also apply in curved space, at least in a small area around the interesting part of space-time. However, this means that the derivative of the energy-momentum tensor must be zero:

$$\nabla^\mu T_{\mu\nu} = 0$$

This would also mean that the derivative of the Ricci tensor would have to be zero, but this will not be the case in every geometry. However, from the Bianchi identity, one obtained that the derivative of a modified Ricci tensor is zero, namely the derivative of the Einstein tensor. With this we have:

$$G_{\mu\nu} = R_{\mu\nu} - \frac{1}{2} R g_{\mu\nu} = \kappa T_{\mu\nu}$$

This is in principle the basic equation of general relativity. However, the constant of proportionality $\varkappa$ is still unknown. We receive this, however, if we consider the boundary case of the Newtonian mechanics.

The rest energy ϱ is the predominant contribution to the energy-momentum tensor for particles moving slowly (Newtonian limit case). It is in the 00 component of the tensor, so we have:

$$\rho = T_{00}$$

The distortion of the space is approximated in first approximation by a straight line. One has then as remaining components of the field equation:

$$\nabla^2 h_{00} = \kappa T_{00}$$

In classical mechanics, the Poisson equation applies to a Newtonian potential:

$$\nabla^2 \Phi = 4\pi G \rho$$

In three dimensions this equation provides the potential:

$$\Phi(r) = \frac{-Gm}{r}$$

As we have (see above):

$$h_{00} = -2\Phi$$

and considering that formulas have so far been derived for c=1, we finally get the Einstein equation in all its beauty:

$$G_{\mu\nu} = R_{\mu\nu} - \frac{1}{2}Rg_{\mu\nu} = \frac{8\pi G}{c^4}T_{\mu\nu}$$

Literature

Aharonov 1959: Yakir Aharonov, David Bohm: *Significance of Electromagnetic Potentials in the Quantum Theory.* In: *The Physical Review.* 115, Nr. 3, 1959, S. 485–491.

Aharonov 1988: Yakir Aharonov et al.: *How the result of a measurement of a component of the spin of a spin-1/2 particle can turn out to be 100.* In: *Physical review letters* 60.14 (1988), S. 1351.

Aharonov 2013: Yakir Aharonov et al.: *Quantum cheshire cats.* In: *New Journal of Physics* 15.11 (2013), S. 113015.

Aharonov 2017: Yakir Aharonov, Eliahu Cohen, Fabrizio Colombo, Tomer Landsberger, Irene Sabadini, Daniele C. Struppa, Jeff Tollaksen: *Finally making sense of the double-slit experiment.* In: *Proceedings of the National Academy of Sciences*, 114(25), 2017, S. 6480-6485.

Aspect 1982: Alain Aspect, Philippe Grangier, G. Roger: *Experimental Realization of Einstein-Podolsky-Rosen-Bohm Gedankenexperiment: A New Violation of Bell's Inequalities*, In: *Physical Review Letters*, Band 49, 1982, S. 91–94,

Bell 1964: John Stewart Bell: *On the Einstein Podolsky Rosen paradox.* In: Physics, 1964, 1(3), S. 195 – 200.

Bennett 1993: Charles Bennett *et al.*: *Teleporting an unknown quantum state via dual classical and Einstein-Podolsky-Rosen channels.* In: *Phys. Rev. Lett.* 70, 1993, S. 1895.

Bohm 1952a: David Bohm: *A suggested interpretation of the quantum theory in terms of hidden variables I.* In: *Phys. Rev.* 85, 166 (1952).

Bohm 1952b: David Bohm: *A suggested interpretation of the quantum theory in terms of hidden variables II.* In: *Phys. Rev.* 85, 180, (1952).

Bohm 1957: *David Bohm; Y. Aharonov: Discussion of Experimental Proof for the Paradox of Einstein, Rosen, and Podolsky. In: Physical Review, 108, 1957, S. 1070.*

Bohr 1913: Niels Bohr: *I. On the constitution of atoms and molecules.* In: *The London, Edinburgh, and Dublin Philosophical Magazine and Journal of Science.* 26 (151), 1913, S. 1–25.

Boltzmann 1877: Ludwig Boltzmann: *Über die Beziehung zwischen dem zweiten Hauptsatz der mechanischen Wärmetheorie und der*

Wahrscheinlichkeitsrechnung respektive den Sätzen über das Wärmegleichgewicht. In: *Sitzungsber. d. k. Akad. der Wissenschaften zu Wien* II 76, S. 428 (1877). Nachdruck in *Wissenschaftliche Abhandlungen von Ludwig Boltzmann*, Band II., S. 164–223.

Bouwmeester 1997: Dik Bouwmeester et al.: *Experimental Quantum Teleportation*, in: *Nature* 390, 1997, S. 575–579.

Casimir 1948: Hendrik Casimir: *On the attraction between two perfectly conducting plates.* In: *Proc. Kon. Nederland. Akad. Wetensch.*, B51, 1948, S. 793.

Clauser 1969: Clauser, John F., et al.: *Proposed experiment to test local hidden-variable theories. Physical review letters*, 23. Jg., Nr. 15, 1969, S. 880.

Compton 1923: Arthur H. Compton: *A Quantum Theory of the Scattering of X-rays by Light Elements.* In: *Physical Review.* Band 21, Nr. 5, 1923, S. 483–502.

Davisson Germer 1927: Davisson, C. and Germer, L. H.: *Diffraction of Electrons by a Crystal of Nickel.* In: *Phys. Rev.* Band 30, 1927, S. 705–740.

De Broglie 1925: Louis de Broglie: *Recherches sur la théorie des quanta*, Doktorarbeit, Paris, 1924. In: *Ann. de Physique* (10) **3**, 1925, S. 22.

De Broglie 1927: Louis de Broglie: *La mécanique ondulatoire et la structure atomique de la matière et du rayonnement.* In: *J. Phys. Radium*, 1927, 8 (5), S. 225-241.

Dürr 1992: Detlef Dürr, Sheldon Goldstein, Nino Zanghi: *Quantum equilibrium and the origin of absolute uncertainty.* In: *Journal of Statistical Physics*, 67. Jg., Nr. 5, 1992, S. 843-907.

Dürr 1993: Detlef Dürr, Sheldeon Goldstein, Nino Zanghi: *A global equilibrium as the foundation of quantum randomness.* In: *Foundations of Physics*, 23. Jg., Nr. 5, 1993, S. 721-738.

Einstein 1905a: Albert Einstein: *Zur Elektrodynamik bewegter Körper.* In: *Annalen der Physik und Chemie.* 17, 1905, S. 891–921.

Einstein 1905b: Albert Einstein: *Ist die Trägheit eines Körpers von seinem Energieinhalt abhängig?* In: *Annalen der Physik.* Band 323, Nr. 13, 1905, S. 639–643.

Einstein 1905c: Albert Einstein: *Ueber einen die Erzeugung und Verwandlung des Lichtes betreffenden heuristischen Gesichtspunkt.* In: *Annalen der Physik.* 322, Nr. 6, 1905, S. 132–148.

Einstein 1916: Albert Einstein: *Die Grundlage der allgemeinen Relativitätstheorie*. In: *Annalen der Physik*. Band 354, Nr. 7, 1916, S. 769–822.

Einstein 1935: Albert Einstein; Boris Podolsky; Nanthan Rosen: *Can Quantum-Mechanical Description of Physical Reality be Considered Complete?* In: *Physical Review*, 47 (10), 1935: S. 777–780.

Freedman 1972: Freedman, Stuart J.; Clauser, John F.: *Experimental Test of Local Hidden-Variable Theories*. In: *Physical Review Letters*, 28 (14), 1972, S. 938-941.

Gamow 1966: George Gamow: *Thirty years that shook physics*, Dover Publication, 1966.

Giustina 2015: Marissa Giustina, Marissa: *Significant-loophole-free test of Bell's theorem with entangled photons*. In: *Physical review letters,* 115.25 (2015), S. 250401.

Heisenberg 1925: Werner Heisenberg: *Quantentheoretische Umdeutung kinematischer und mechanischer Beziehungen*. In: *Zeitschrift für Physik*. Band 33, 1925, S. 879.

Heisenberg 1927: Werner Heisenberg: *Anschaulicher Inhalt der quantenmechanischen Kinematik*. In: *Zeitschrift für Physik*, Band 43, 1927, S. 172.

Jacques 2007: Vincent Jacques: *Experimental Realization of Wheeler's Delayed-Choice Gedanken Experiment*. In: *Science* 2007, 315 (5814). S. 966–968.

Jaffe 2005: Robert Jaffe: *Casimir effect and the quantum vacuum*. In: *Physical Review D*, 72 (2), 2005, S. 021301.

Jönsson 1974: Claus Jönsson: *Electron Diffraction at Multiple Slits*. In: *American Journal of Physics*. Band 42, 1974, S. 4–11.

Kim 2020: Yosep Kim et al.: *Observing the quantum Cheshire cat effect with noninvasive weak measurement*. In: *npj Quantum Information* 7.1 (2021), S. 1-6.

Lamoreaux 1997: S. K. Lamoreaux, S. K.: *Demonstration of the Casimir Force in the 0.6 to 6 μm Range*. In: *Physical Review Letters*, 78 (1), 1997, S. 5–8.

Lemmel 2022: Lemmel, Hartmut, et al.: *Quantifying the presence of a neutron in the paths of an interferometer*. In: *Physical Review Research*, 2022, 4. Jg., Nr. 2, S. 023075.

Merli 1976: P G Merli, G F Missiroli, G Pozzi: *On the statistical aspect of electron interference phenomena*. In: *American Journal of Physics*, 1976 44 (3): 306–307.

Minkowski 1907: Hermann Minkowski: *Das Relativitätsprinzip*. In: *Annalen der Physik*. 352, Nr. 15, S. 927–938.

Philippidies 1979: C. Philippidis, C, Dewdney, B. J. Hiley: *Quantum interference and the quantum potential*. In: *Il Nuovo Cimento B (1971-1996), 52*(1), 1979, S. 15-28.

Planck 1900a: Max Planck: *Über eine Verbesserung der Wien'schen Spectralgleichung*. In: *Verhandlungen der Deutschen Physikalischen Gesellschaft*, 2: 202–204.

Planck 1900b: Max Planck: *Zur Theorie des Gesetzes der Energieverteilung im Normalspectrum*. In: *Verhandlungen der Deutschen Physikalischen Gesellschaft*, 2: 237–245.

Rutherford 1911: E. Rutherford, *The Scattering of a and β Particles by Matter and the Structure of the Atom*. In: *Phil. Mag*. 6, vol. 21, 669–688 (1911).

Schrödinger 1926a: Erwin Schrödinger: *Quantisierung als Eigenwertproblem*. In: *Annalen der Physik*. Bd. 79, 1926, S. 361.

Schrödinger 1926b: Ernst Schrödinger: *An Undulatory Theory of the Mechanics of Atoms and Molecules*. In: *Physical Review*. 28 (6), 1926, S. 1049–1070.

Schrödinger 1935a: Ernst Schrödinger: *Discussion of probability relations between separated systems*. In: *Mathematical Proceedings of the Cambridge Philosophical Society*. 31 (4), 1935, S. 555–563.

Schrödinger 1935b: Ernst Schrödinger: *Die gegenwärtige Situation in der Quantenmechanik*. In: *Die Naturwissenschaften*, 1935, 48, S. 807 – 812, 49, S. 823 – 828, 50, S. 844 – 849.

Sommerfeld 1916a, b: Arnold Sommerfeld: *Zur Quantentheorie der Spectrallinien (I + II)*. In: *Annalen der Physik*. 51, 1916, S. 1–94 und 125-167.

Tonomura 1986: Akira Tonomura et al.: *Evidence for Aharonov-Bohm effect with magnetic field completely shielded from electron wave*. In: *Physical review letters* 56.8 (1986), S. 792.

Wheeler 1978: John Archibald Wheeler: *The "past" and the "delayed-choice" double-slit experiment*. In: *Mathematical foundations of quantum theory*. Academic Press, 1978. S. 9-48.

Wheeler, 1983: John Archibald Wheeler: *Law without law*. In: John Archibald Wheeler and Wojciech Hubert Zurek (Editors), *Quantum Theory and Measurement*, Princeton University Press, 1983, S. 182 – 213.

Wigner 1961: Eugene P. Wigner: *Remarks on the Mind-Body Question*. In: Good, I. J. (ed.), *The Scientist Speculates: An Anthology of Partly-Baked Ideas*. London: Heinemann, 1961.

List of figures

S. 74: https://de.wikipedia.org/wiki/Datei:Jupiter_by_Cassini-Huygens.jpg,
https://de.wikipedia.org/wiki/Datei:The_Blue_Marble_(remastered).jpg
S. 82: Aspect 1982
S. 109:
http://hubblesite.org/newscenter/archive/releases/1990/20/image/a/
S. 114:
https://de.wikipedia.org/wiki/Datei:Th%C3%A9orie_des_cordes-%C3%A9chelle.PNG
S. 116: https://de.wikipedia.org/wiki/Datei:M-Theory.svg
S. 123: Aharonov 2013
S. 125: Lemmel 2022

www.ingramcontent.com/pod-product-compliance
Lightning Source LLC
Chambersburg PA
CBHW060838220526
45466CB00003B/1154